Integrated Services Digital Networks

Integrated Services Digital Networks

from concept to application

2nd edition

John Ronayne

CRC Press
Taylor & Francis Group
Boca Raton London New York

CRC Press is an imprint of the
Taylor & Francis Group, an **informa** business

CRC Press
Taylor & Francis Group
6000 Broken Sound Parkway NW, Suite 300
Boca Raton, FL 33487-2742

First issued in hardback 2017

© 1996 by John Ronayne
CRC Press is an imprint of Taylor & Francis Group, an informa business

No claim to original U.S. Government works

ISBN-13: 978-1-8572-8406-5 (pbk)
ISBN-13: 978-1-1384-7244-0 (hbk)

**Visit the Taylor & Francis Web site at
http://www.taylorandfrancis.com**

**and the CRC Press Web site
http://www.crcpress.com**

British Library Cataloguing in Publication Data
A CIP record for this book is available from the British Library.

Typeset in Sabon and Univers.
Printed and bound by
Antony Rowe Ltd, Eastbourne.

Contents

Preface

In the preface to the first edition I foretold that readers would be consulting a second edition in some two years' time. Like all prophecies in the telecommunications and electronics field, this one has turned out to be both right and wrong. It was right in that a second edition was certainly needed after two years, although for reasons other than those I had predicted, and wrong in that the promised second edition has in fact taken eight years to materialize and covers some radically different topics, partly because of this additional unpredicted delay.

In 1987, when most of the first edition was written, it was reasonable to expect that the ISDN would be implemented widely, at least in the western world, within two years and that the controversies and difficulties outlined in Chapters 6 and 7 of the first edition would be resolved. Thus, the new edition would be required to describe actual implementations and definitive resolution of difficulties.

In the event, the nature of the world has changed, "west" and "east" now have different meanings, and widespread implementation has only recently started or is just about to start (UK, most of Europe, Japan, USA, France and Germany). International interconnection of ISDNs is also in its early stages. CCITT recommendations and related national standards are now complete to the extent that national and international interworking is possible and ISDN terminal equipment can be designed, manufactured, tested and installed with some degree of confidence. Nevertheless, gaps remaining in the recommendations continue to inhibit easy implementation and international interworking.

Because the ISDN is now, at last, being implemented, many more readers will be looking for an appropriate book on the subject and certainly, among the many more books that are now available, will not choose one written so long ago. For this reason if for no other a new edition is necessary. More seriously, however, although much of the material in the first edition is still relevant and requires additions rather than changes, a new edition is essential to detail the actual implementations now being put in place. Although development and ISDN deployment has moved more slowly than expected since the first edition, technology, as always, has raced ahead. Thus, ISDN implementation is occurring in an environment that is already changed and promises to be radically different from that envisaged in the first edition.

In the first edition the references to B-ISDN were brief and, for the most part, scathing. Like it or not, B-ISDN is now a fact, although again in rather a different realization than its proponents were predicting in 1987. It is necessary now to devote considerably more space to the subject and to the enabling technologies of the B-ISDN, the synchronous digital hierarchy (SDH) and asynchronous transfer mode (ATM). It is satisfying to the author that the realization of the B-ISDN using these technologies borrows some of the insights provided in the systems design proposal provided in Chapter 8 of the first edition.

In this second edition, therefore, the main changes and additions are as follows.

(a) All figures from and references to the CCITT Blue Book recommendations ratified in 1988 or to their successors in the "white books" have been updated. In particular, the new editions of the main ISDN protocol standards, ITU-T Recommendations Q.920/921 and Q.931, have been taken into account.

(b) Information on ISDN implementations worldwide, with special reference to the problems and their solution, has been added. The accounts of the problems of the ISDN outlined in Chapters 6 and 7 (Chapters 8 and 9 of the second edition) have been expanded.

(c) The account of the standards setting environment has been expanded to include European developments (ETSI was not born when the first edition was published).

(d) Three new chapters on the B-ISDN and the relationship between "narrow band ISDN" and its new, high-capacity fellow have been included.

(e) The telecommunications network environment into which the ISDN is being introduced, including the widespread use of fibre, high-capacity transmission using the synchronous digital hierarchy and a new perception of the place and nature of the telephone (now more properly telecommunications) exchange, is given new treatment.

(f) The reference list has been updated to include the many other books on the ISDN now available and much new reference material expanding the accounts included in this volume.

Note that since the first edition the ITU has reorganized itself so that CCITT recommendations are now referred to as ITU-T Recommendations.

John Ronayne
Nailsea, March 1996

Reference citations
Most references are cited in superscript thus, "[XX]". Direct references in the text are referred to by the reference number, "X", without brackets. The listed references start on page 261, and other useful references not cited in the text are included in Further Reading on page 267.

Preface to the first edition

Writing a book on any technically advancing subject has similarities to aiming at a moving target. A book about integrated services digital networks is fated to be out of date before it is printed. The author is conscious of matters discussed in Chapters 6 and 7 of which he was unaware when writing Chapters 2 and 3. The book is therefore offered to those many readers who need to know about the ISDN now, with the caution (and the implied promise) that they will need to consult the second edition in some two years' time.

The exciting part of writing about a fast-moving technology is the possibility that what one writes may not be without influence on the path of development. With this in mind, I have not always desisted from providing opinion as well as technical fact.

The difficult part of such a project is the impossibility of mastering, or even knowing about, all the movements taking place in the technology. For much of the information and not a few of the ideas presented, I have to thank many colleagues and friends. Chief among these, I feel bound to single out the following: Hans Diemel, Peter Dudley and Mike Thomas of Northern Telecom; Geoff Harland of STC Technology Ltd; David Colbeck of Thorn Ericsson; Vic Teacher of STC plc; Arthur Orbell and Chris Wood of British Telecom, who have patiently read my manuscript as well as providing much help and advice; Brian Terry and all the members of his group at BT for many helpful discussions; and my wife Mary, who has, again, not only typed the manuscript but straightened out much convoluted English prose. To these, and to many others, I am indebted for much of the accuracy and truth of my account. For the opinions expressed and for any mistakes and omissions, I alone am responsible.

John Ronayne
Nailsea, March 1987

CHAPTER 1

The ISDN objective

1.1 Introduction

The abbreviation for integrated services digital network, ISDN, was introduced in about 1979 to represent the objective towards which digital telecommunications development was tending. The ISDN is defined as follows.

> An ISDN is a network, in general evolving from a telephony IDN, that provides end-to-end digital connectivity to support a wide range of services, including voice and non-voice services, to which users have access by a limited set of standard multipurpose user–network interfaces.
>
> CCITT Red Book, Fascicle III.5, p.3.

What is intended by this definition is illustrated in Figure 1.1 in terms of a future residential subscriber to the telephone service who uses the ISDN. As at present, the subscriber is connected to the network by a pair of wires. As at present, a couple of telephone extensions are fitted. There the similarity ends. This ISDN subscriber also possesses a personal computer (PC) on which are used software programs (e.g. games), ordered, paid for and obtained from a software retailer via the telephone line. Perhaps, too, the subscriber performs business tasks on the PC, the results of which are accessed by colleagues from their own home PCs. Our subscriber also subscribes to a viewdata service, receiving the viewdata frames either on the PC or on the television. The subscriber's gas, electricity and water meters are read by the relevant service companies using the telephone network for access and the resulting bill is presented to the subscriber via a viewdata message. From home the subscriber can pay bills and purchase services and goods using electronic funds transfer. This may require the use of a credit card in a special reader or it may be performed by a secure coding and password system. The kitchen of the ISDN home is equipped with gadgets to control the cooker and the central heating, allowing the absent family to command the house to be warm and the dinner cooked on its return, and while they are absent the home security system can have access via ISDN to the police or security firm.

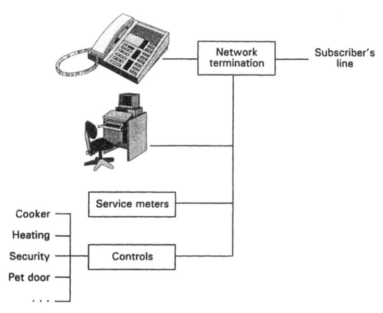

Figure 1.1 The ISDN home.

The essential point about all this is that only the equivalent of one pair of wires connects the customer's terminal to the network. The subscriber may speak on the telephone at the same time as the PC is loading new data and the service company is reading the meter, all over the single connection.

It may well be that the number of private homes desiring this degree of sophistication will be a small proportion for very many years to come. It is undoubted, however, that an equivalent degree of sophistication is already available in many offices and factories, often accompanied by a wonderful tangle of wires and cables. Figure 1.2 shows a typical secretary's desk today: telephone connected by a two-wire line to the PABX; word processor connected by coaxial cable to a local area network (LAN), giving access to the word processing computer and the company database; facsimile machine using another telephone pair; and telex, which, in fairness to modern practice, is assumed to be incorporated in the PC but uses yet another telephone pair connected to a different network. Elsewhere in the office further systems are in use: computer-aided design, viewdata, voice messaging (the secretary of Figure 1.2 may have this with no extra complication), video security, videoconferencing, electronic funds transfer, etc. All, or most, will at present require separate network connections.

With the ISDN, all or most of these services can be provided to the user (the secretary in Figure 1.2) over a single pair of wires. Moreover, several of the services can be in communication simultaneously. At present, most of the services described are provided over separate private and public networks. With integration the network providers and the service providers are free to decide

Figure 1.2 The "office automated" desk.

to use integrated networks to provide all or most of the services.

Furthermore, the ISDN basic access can support up to eight voice or data terminals so that the single ISDN access can perform the functions previously requiring a small PABX or key system *and* a small LAN. In many cases users will find this to be the most cost-effective means of providing such facilities even without taking into account the facilities and services made available by the ISDN itself.

1.2 Historical review

It will be helpful to understand the requirements of all the various communications services that are candidates for integration. To gain this understanding we will review the historical development of telecommunications.

1.2.1 Telegraph

The telegraph was the first electrical communications system to be introduced. Using direct current pulse signals on single-wire earth-return lines or, later, two-wire lines, it was possible to detect such signals over long distances by using sensitive galvanometers as the detectors. Transatlantic telegraph cables were in use by the mid-1800s. Pulse codes such as Morse code were devised and highly skilled operators could achieve speeds of perhaps 30 words per minute. Apart from the limits of human skill, speed restrictions were set by the long transmission delays over long circuits.

The relative slowness of telegraphic transmission distorted human communication as all unnecessary words were excluded. A new form of language, "telegraphese", was developed that can still be discerned in use, particularly in telex messages, long after its utility has disappeared.

The telegraph introduced to communications the use of simple digital codes and used protocols to ensure the accuracy, privacy and correct addressing and reception of the messages. Protocols were already highly developed in other spheres, notably naval communications, where quite complex messages were conveyed by coloured flags, visible in the heat and smoke of battle but indecipherable to the enemy unaware of the protocol.

1.2.2 Telephone

The invention of the telephone and its rapid proliferation in the US transformed telecommunications. Now the users could communicate directly, apparently without protocol and without the intervention of skilled operators sending, detecting and recording the messages. The absence of skilled operators urged on the improvement of transmission media to enable untutored users to communicate comfortably over longer distances. Automatic switching and long-distance dialling increased the ease and range of communications still further. Truly long-distance telephony (across the Atlantic, for example) was for long impeded by the limitations of attenuation, bandwidth, interference and channel capacity of radio, the only transmission medium available. Introduction of submarine coaxial cable and repeaters and, later, satellite communications transformed long-distance telephony in both performance and economics.

1.2.3 Telex

The development and manufacture of complex machines, notably in the textile industry, had prepared the way for the production of delicate mechanisms for telephone switching. At about the time that Alexander Graham Bell was promoting the telephone, the typewriter, invented by Christopher Sholes in 1867, was appearing in offices. It is a fairly small conceptual step to equip the telegraph operator with an electric typewriter that sends telegraph code for each letter and digit. The existing telegraph network was first equipped with printing telegraph machines and there was a surprisingly long pause before a dedicated telex network appeared in the 1930s and 1940s, interconnecting telex users directly without the need for telegraph operators.

Perhaps some of the delay in introducing a telex network was caused by international acceptance of the necessary protocols to enable a machine to be called and "answer back" with its identity. The various telex network proposals also tried to provide the facilities of the telegraph office to store and forward messages, broadcast messages, etc.

The telex network and telex machines equipped with punched tape perforators and transmitters first introduced the protocols necessary for direct machine-to-machine communications without human intervention.

The provision of machine-to-machine communication with protocols for answer back confirming identity has led the telex message to be accorded the status of a legally recognized confirmed message. Thus, despite modern improvements in printed word communications (facsimile, viewdata, etc.) the telex facility persists because of its unique legal status and worldwide availability.

1.2.4 Data networks

Strictly speaking, both the telegraph and telex networks are data networks. The term became current when the increasing power of digital computers encouraged the concept of time-sharing tasks requested by many terminals on the same computer. From dedicated star networks within an office or factory, data networks have grown to worldwide transport networks interconnecting computer mainframes with compatible terminals and with other compatible mainframes and providing connectivity between diverse users of personal and small business machines. Today some of this data traffic is carried on the public switched telephone network (PSTN) by converting the binary digital signals into tone signals within the speech band (300–3400 Hz). The speech bandwidth of the public telephone network is too constricted to allow efficient communication between mainframes, which requires speeds up to 19.2 kbit/s. Thus, dedicated private and public data circuits and networks have proliferated.

The drawbacks and difficulties of machine-to-machine communication and the need for strict protocols have become a matter of public awareness. Films such as *War games*, TV series such as *Bird of prey* and many more that have appeared since and the activities of a new species of criminal (the more innocent among them known as "hackers") have made us well aware of the possibilities of fooling the machine into believing that unauthorized access is permitted. This public awareness enhances the difficulty of integrating public and private facilities – the honest among us fear the security dangers, the less honest contemplate methods of infiltration.

1.2.5 Packet switching

The means of communication discussed so far have assumed the existence of a dedicated link or a circuit switched connection between the communicating persons or machines throughout the "conversation". Data communication has characteristics that are very different to those of voice. For voice, a two-way path must be permanently available, if only for the comfort of the speaker as dialogue tends towards monologue. Data communication does not necessarily require simultaneous both-way conversation. Moreover, data communication has different patterns to voice, ranging from a very rapid query–pause–response sequence to a short query followed by a very long response as programs or data are downloaded, for example. A further data mode might be a continuously open connection that is occasionally used for a short burst of data. The data may also be subjected to error control procedures. The data connection therefore must be set up much faster than voice (so as not to waste the time of a valuable mainframe) but may then be used for quite small proportions of the total connection time.

For data communications, therefore, a totally different method of connection can be used. Rather than retaining a dedicated communications path for the duration of the call (circuit switching) and having a rather leisurely protocol in setting up this path, a method of message switching can be used. For message switching the total data conversation can be broken up into elements, and each element will be sent separately involving, if necessary, the rapid set-up of the appropriate path just for that element. This form of message switching is called packet switching.

Figure 1.3 illustrates the two concepts of circuit and message switching. It can be seen that each element of the message must be accompanied by labels indicating the source and destination. It is evident also that messages must be stored at the various switching and controlling nodes.

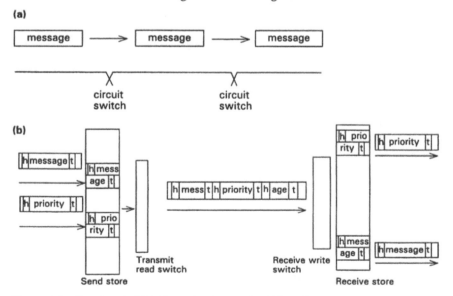

Figure 1.3 Circuit switching (a) and message switching (b).

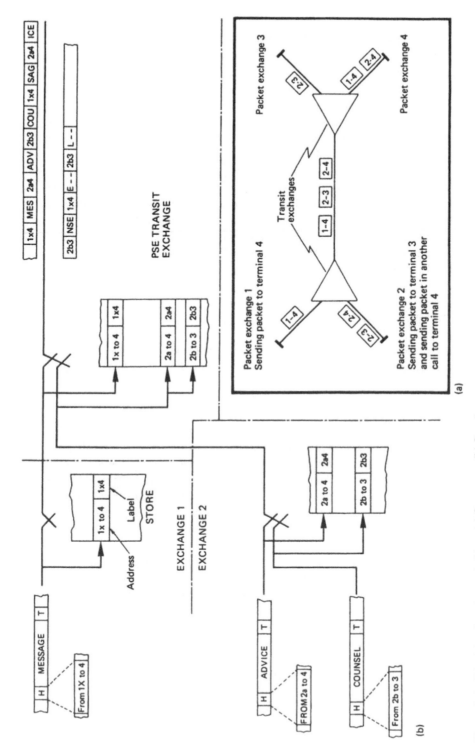

Figure 1.4 Principle of packet switching. (a) Summary. (b) Detail.

The need to repeat the address information with every portion of a message represents a significant overhead for any but the very basic message switching system shown in Figure 1.3. The solution, adopted by the various packet-switching protocols in existence, is to identify and store the address information against a label at each node so that only the initial packet carries the full address and subsequent packets carry only the identifying label. The only facility held throughout the duration of the call is the address and label stored at each node (Fig. 1.4).

Packet-switching networks have been provided by various public administrations, for example the BT network, known as PSS, was introduced in the early 1980s. Some are extended internationally. Ethernet,[1] for instance, extended across Europe and could be accessed by PSS and other national networks. Many other packet-switching networks exist, serving government agencies and private users. Arpanet, one of the first to be established, serves the US Defense Department.

1.2.6 Other services

There are other telecommunications services that may also require separate networks but which may, in some cases, be provided by one of the networks already described. Such services include:

1.2.6.1 Radio telephone
Figure 1.5 shows the common divisions into which the subject of mobile communications can be grouped. Communications with ships and aircraft have been omitted. All these services, apart from paging, have been introduced since 1975. This is the underlying reason for the confusing number of different initiatives. Complementary and competing services have been introduced to the market at substantially the same time. There has therefore been a confusion of overlapping and competitive developments. A short description of each of the main services is provided here.

(a) *Paging systems* permit users, who carry a pocket receiver, to be alerted to the fact that they are wanted. In most cases it is necessary for the user to contact a central point in order to find out what is wanted.

(b) *Private mobile radio (PMR)* provides two-way voice and, possibly, data communication between the user and a central point or between users belonging to the same closed user group.

(c) *Cellular systems* provide continuous communication between mobile users and any other telecommunications terminal connected to the worldwide public network, whether mobile or fixed. They are named after the technology of providing small transmitter/receivers serving cells

---

1. An early, publicly available, packet-switching network, not to be confused with the Ethernet LAN protocol (see Ch. 2).

so that the scarce resource of radio frequencies can be reused in non-adjacent cells. The moving mobile is handed on to the cell best placed to continue the communication.

(d) *Cordless systems* started with the desire to take the telephone into the garden or the bath. The cordless phone communicates with a base station, which can be in the home or, with the use of personal communications networks (PCNs), in a public place.

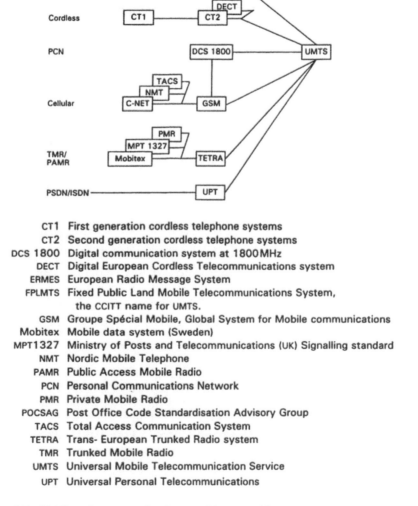

CT1 First generation cordless telephone systems
CT2 Second generation cordless telephone systems
DCS 1800 Digital communication system at 1800 MHz
DECT Digital European Cordless Telecommunications system
ERMES European Radio Message System
FPLMTS Fixed Public Land Mobile Telecommunications System,
    the CCITT name for UMTS.
GSM Groupe Spécial Mobile, Global System for Mobile communications
Mobitex Mobile data system (Sweden)
MPT1327 Ministry of Posts and Telecommunications (UK) Signalling standard
NMT Nordic Mobile Telephone
PAMR Public Access Mobile Radio
PCN Personal Communications Network
PMR Private Mobile Radio
POCSAG Post Office Code Standardisation Advisory Group
TACS Total Access Communication System
TETRA Trans- European Trunked Radio system
TMR Trunked Mobile Radio
UMTS Universal Mobile Telecommunication Service
UPT Universal Personal Telecommunications

Figure 1.5 Mobile radio system development history and future.

Eventually there will probably be a merging of cordless systems and cellular systems. Certainly, the use of words such as "total" and "universal" in the titles of Figure 1.5 is not yet representative of fact because of the variety of systems. It is, however, envisaged that PCNs will permit developments that will satisfy the needs of cordless telephony and cellular mobile telephony.

1.2.6.2 Telemetry and teleconnect
These are low data rate services to read service meters, operate devices, extend alarms, etc.

1.2.6.3 Facsimile
This is essentially a service to read and encode documents and pictures, transmit them and reconstitute them at the distant terminal.

1.2.6.4 Teleconferencing
This service provides conference conversation between any reasonable number (say 20) of PSTN users anywhere in the world.

1.2.6.5 Viewdata or videotex
In the UK, the viewdata service that BT used to provide was known as Prestel. It was a long-drawn-out failure not because of inadequacies in the service itself, but through definite inadequacies in the commercial arrangements at its launch and throughout its life. Similar services, notably Minitel in France, continue to be highly successful. Such a service allows access over the PSTN to computer databases containing information supplied by others either for access by the general public (airline timetables, for example) or for access by closed user groups (finance market information, for example). The user can communicate with the database, but usually only in a limited way, to obtain information, order a ticket or leave a message in a "letterbox" for example. Users who are information providers can communicate more generally with the database to set-up new "frames" of information or to change existing frames. The term used by the International Telecommunications Union (ITU) is videotex.

1.2.6.6 Slow-scan video
A full video facility cannot be provided on the bandwidths normally allowed for telephone use. Slow-scan TV is often adequate for monitoring black and white or, even, colour pictures from, for example, security cameras. A pioneering use of slow-scan video was introduced by Frenchay Hospital in Bristol to allow consultants to examine X-ray pictures transmitted to them over the PSTN.

1.2.6.7 Teletex

This is similar perhaps to telex; a teletex service will allow full both-way textual communication between terminals. This could be used for communicating between word processors or as a form of electronic mail.

1.2.6.8 Teletext

This is a service of information continuously broadcast by TV and selected locally by the user. In the UK, Ceefax and Oracle are teletext services. (Note the unnecessary confusion created by using similar names for different services.)

1.2.6.9 Telemail

This is a more widespread, cheaper form of teletext, provided perhaps as storage of textual information with a notice of receipt displayed on the user's TV screen. The user views the text on the TV or obtains a print from a simple printer.

1.2.6.10 Home newspapers

This could be part of a viewdata service. TV receivers with local printing facilities have appeared on the market.

1.2.6.11 Electronic funds transfer (EFT)

Associated with viewdata or communications services, EFT allows the user to order and pay for goods and services using a credit card reader or other form of secure access and validation arrangement. A particular form of EFT is EFT at the point of sale (EFTPOS), allowing the credit limit check associated with the use of credit cards to be combined with the actual transaction.

1.2.6.12 Wideband services

Video at normal speeds and in colour, colour facsimile, videoconferencing and viewphone are all services that are technically possible but which require wider transmission and switching bandwidths than that available with normal telephony.

1.2.7 Integration

The existence of separate networks for separate services tends to ease the problems of providing the service. To the user, however, the provision of the service over a separate network is a disadvantage. Access is available only to other subscribers to the same service unless there are internetwork gateways. In addition, the administration of each separate network represents a separate and considerable overhead that, with integration, could be shared over the several integrated services. As an illustration, UK telephone subscribers to

Telecom Gold, a teletex service, can gain access to the telex network, but only by using quite complex access protocols and at the expense of the telex answer back feature, which does not work over Telecom Gold to the originating user. Thus, a message thought to be legally binding may be sent with no confirmation of receipt. Telecom Gold subscribers, therefore, have access only to other Gold subscribers and to users of compatible teletex services, but only with some difficulty and diminished facilities to the worldwide telex service.

A second illustration is the telex service itself. Until 1952, a telex service was provided in the UK using the PSTN, but it was difficult to provide a sufficiently good-quality service by this means. In that year the telex service was transferred to a separate network and has provided dubious economic returns ever since.

There are therefore considerable advantages to both the user and the network provider of integrating diverse services over the same telecommunications network. In addition, some of the very real practical difficulties are apparent. Separate networks exist and have a long unexpired economic life. (The UK PSS network has only relatively recently been fully deployed.)

Considerable complication can be foreseen in organizing the charging for many services over the same network. (The charging arrangements for Prestel users and information providers illustrate the complexity.) Solutions to all the technical problems are in their infancy, and some of the practical problems have hardly been considered. Nevertheless, users will insist on integration over their own private networks if nothing else exists so that the ISDN will certainly be introduced. It is important that its introduction is controlled and co-ordinated in a sensible manner so that no part of the existing worldwide compatibility over the existing separate networks is lost.

1.3 The ISDN objective

The objective, then, of the ISDN is to provide the user with easy access to a multiplicity of services over a single connection to a single network. The least important element of the title is the word "digital", although it is the digital network that has simplified the provision of the integrated services.

To meet this objective the discussion so far has indicated some of the requirements. These are now discussed in more detail.

1.3.1 In the local network

The local network[1] is the connection between subscribers and the local exchange. It has the following requirements:
   (a) duplex operation of simultaneous speech, data, telemetry and signalling;

(b) transparency, i.e. no constraints on the message content used in the various services;

(c) operation over two-wire line plant of conventional length and characteristics without regeneration (typical line length is up to 5 km);

(d) extended area working using regenerators;

(e) negligible power consumption in the dormant state but rapid wake-up and alignment to minimize call set-up times (100 ms from "off-hook" to first message);

(f) rapid call set-up times – overall from end to end of the network a time of 1–2 seconds is required compared with 20–30 seconds common in existing telephony networks;

(g) transmission and detection of bit and frame timing signals in both directions;

(h) good, error-free performance (a mean bit error rate target of 1 in 10^7);

(i) no interference to or from conventional transmission and signalling systems carried in the same cable, which should be achieved without the need to reduce the fill (pairs actually used) of cables to accommodate digital line systems up to 2.048 Mbit/s;

(j) connection of nominal 25 mA DC or its equivalent during the active state to provide power to conventional telephones;

(k) monitoring and testing features for fault location and diagnosis;

(l) operation within the normal range of ambient temperature and humidity without the need for air conditioning;

(m) interface compatibility with other forms of local access transmission including radio links, optical fibre, ring distribution systems, etc.;

(n) protection against power cross-connection faults, lightning, etc.;

(o) simple installations achieved at the customer's end at least without the need for sensitive setting-up and alignment procedures;

(p) remote equipment suitable for independent mounting on premises or for integration within customer terminal equipment;

(q) exchange-end equipment in a form suitable for mounting with or integration in the exchange terminal equipment or remote concentrator or multiplex equipment;

(r) channel capacity at the customer's premises capable of simultaneous speech, data, telemetry and signalling and of expansion to provide several such integrated accesses enabling the services to be extended through the customer's PABX if this is suitable;

(s) the ability to multidrop – one or several ISDN accesses from a total of, say, 15 in a 2.048 Mbit/s system are dropped at each subscriber's premises. (Where, then, is the multidrop equipment to be located? In the road? Or in the subscriber's premises, in which case the subscriber becomes, to an extent, responsible for the public network?)

13

1.3.2 In the public network

The remainder of the network, local trunk, trunk and international interconnections with the local exchanges have the following requirements:
  (a) the capacity to switch independently speech, data, telemetry and signalling channels from a single subscriber location;
  (b) the capacity to extract address information, perform validity checks (on the address at least, possibly on the complete message) and transmit message information to the desired destination.
  (c) the ability to access existing specialized networks and to pass messages to and from such networks and the ability to understand and use the specialized protocols of such networks;
  (d) the ability to identify charges and to allocate charges between the various service providers;
  (e) the ability to provide power feeding for some or all of the customer's equipment from the network switching exchanges, this power feed being able to extend at least to customer's equipment that can switch on local power to the remainder of the customer installation;
  (f) the ability to ensure that quite complex customer equipment is "woken up" and then operates correctly.

1.4 ISDN tools

The ISDN objective has been summarized and the resulting network and terminal requirement has been briefly considered. This leads to the need to summarize the technical methods available for achieving the objective. First among these is stored program control (SPC) of the switching nodes of the network leading to shared control between SPC nodes by means of common channel signalling (CCS). Other tools are the digital network itself, the universal protocols satisfied by open systems interconnection (OSI) and the increasing availability of complex functions in small electronic devices fabricated by large-scale integration (LSI). Inherent in most of these technical methods is the requirement of common standards.

1.4.1 Stored program control (SPC)

SPC was introduced into the telephone exchange with the introduction of the first electronically controlled switching systems (in 1957 in Morris, Illinois, USA). In its present, fairly fully developed form, SPC involves the use of normally quite similar control machines (computers) operating upon software programs that express the diverse switching, signalling and administrative requirements. The software programs operate upon software data describing the configuration details of the exchange to establish and control the network connections required.

The revolution involved in SPC, crucial to the practicability of the ISDN, is the separation of the logic of communications connections from the hardware of the connection process itself. The SPC machine can think, within the limits defined by its software, and can communicate with co-operating SPC machines. The means of communication is another matter requiring CCS, perhaps OSI, for its realization.

In theory, although not always in practice, SPC permits an almost unlimited degree of sophistication to be achieved at the switching node. This sophistication is already used in, for example, packet switching or in the complex "hand-off" protocols necessary to maintain communication with a mobile user moving between different cells of a cellular radio telephone network. In the ISDN this sophistication is required to sort and direct the diverse connections possible from the network terminal (the subscriber). It permits the complex allocation of charges between diverse network and service providers. Sophistication is also necessary to maintain conventional telephone service to users not interested in ISDN offerings (the majority probably for many years to come) and to distinguish these from ISDN users.

1.4.2 Common channel signalling (CCS)

CCS is a method of signalling that concentrates all the management and connection signals relating to each of a multiplicity of communication channels onto a single "common" channel (Fig. 1.6).

CCS was suggested by the existence of high-speed SPC exchange control.

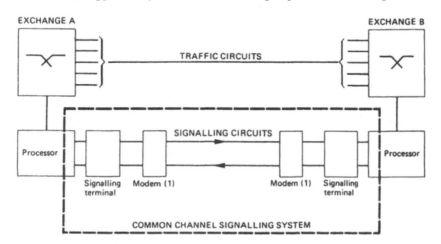

Figure 1.6 Principle of common channel signalling.

Such controlling machines could benefit by co-operation between machines at the various switching nodes. The co-operation required high-speed, secure communication. Why should this high-speed communication link not be used also to carry the control signals associated with each of the many conventional communication channels between the switching nodes[2]?

CCS is essential to the ISDN as it has already provided the communication capacity, speed and protocols to deal with the much more complex requirements of ISDN connections to or between diverse services.

1.4.3 Digital networks

It is the digitization of the telephone network that has been the impetus for the ISDN. The telephone network far outstrips in size and extent any other existing communications network. While the telephone network remains analogue, a channel bandwidth of 300–3400 Hz is provided. Digital data to be transmitted over such a bandwidth must be modulated into analogue form, but the maximum achievable bit rate is only about 2.4 kbit/s if interference with neighbouring channels is to be avoided. Higher bit rates are used over analogue channels, but only by using complex coding techniques. The normal bit rate used for digitally encoded speech is 64 kbit/s, and such a channel capacity will accept many data channels. With a fully digital network or transmission facility, there is no interference problem.

1.4.4 Open systems interconnection (OSI)

OSI is an attempt to rationalize and compartmentalize the conversation between processors and communicating computers. In the case of telephone switching, the advent of SPC necessitated communication between processors, but communication limited by the restricted subject area they talked about – communications switching and signalling. The object was to establish a connection between human users; the conversation thereafter was a matter for the users alone. When the user is a machine then communication becomes more difficult. In the human conversation the network did not perhaps concern itself with language differences between communication users.[2] When the users are machines it is perhaps more important to ensure that a real conversation is in progress and is understood at either end.

The OSI model[3] has been devised to rationalize the conversation between the various communicating processors that may become involved in a (data) communications connection. The basic idea is to define the communications

---

2. All international signalling systems defined by the ITU do in fact provide for a language digit signal to ensure that if the intervention of an operator is requested then the operator will have the correct languages.

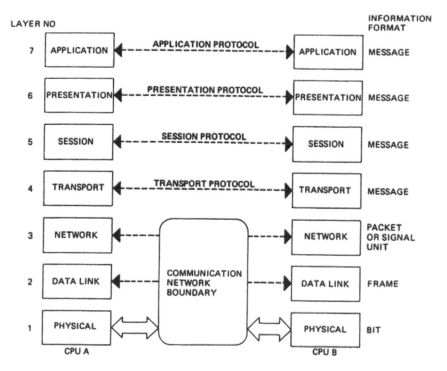

Figure 1.7 ISO Open Systems Interconnection (OSI) reference model.

requirement in terms of layers, each layer contributing a defined modular section of the total communications requirement.

Figure 1.7 shows the OSI seven-layer model, which embodies this definition. The layers are defined as levels of abstraction away from the communications link itself and towards the communicating processor or, more properly, the task requiring the communication. Each layer is defined in terms of its modular activity or purpose (the contribution it makes to the total communication) and in terms of its defined interface with the layer below and the layer above.

Examination of Figure 1. 7 reveals the intention that each layer communicates with its opposite number at the far end of the link oblivious of the contribution of the intervening layers and using a protocol unique to that layer.

The CCS systems already defined for communications, ITU-TS signalling system no. 6 (#6) and, more particularly, ITU-TS signalling system no. 7 (#7), were the first defined communications systems to embody this idea. They used it perhaps even before its formulation by the International Standards Organization (ISO).

17

1.4.5 Large-scale integration (LSI)

LSI, typified by the microprocessor on a chip, is essential to the ISDN because it puts within reach of telephone users both the equipment to provide the digital termination at users' premises and the equipment (processor, facsimile, advanced private telephone exchange, etc.) necessary to make full use of all the features of digital connection. The ISDN depends for its utility upon the intelligent terminal, which is capable of receiving, processing and transmitting messages on demand from a remote user as well as responding to the local users' requests. We have seen these intelligent terminals invade our offices and homes over the past 10–15 years in parallel with the development of LSI. The "dumb terminal" on the computer mainframe was followed by the responsive terminal, dial-up links, the microcomputer, the personal computer, cheap home personal computers, videotext, viewdata and EFT. It is these devices, and more, that will now be given access via the ISDN to similar devices worldwide.

1.4.6 Standards

What has been said already may have sufficiently explained why standards should be listed as a tool. It is very clear that the increased complexity of the communications possibilities and of the communications themselves demand rigid standards for successful universal, integrated connectivity.

The pattern to which communications engineers have become accustomed is that a perceived need has been experimentally applied and that subsequently the standard-setting organizations have, if necessary, embodied the essential definition in a standard. This was the way that the standards were arrived at for conventional international signalling systems (CCITT Rl, R2, #6, #7 and no. 5 are examples[4–8]). For the ISDN this process has been turned on its head, underlining both the increased scale and importance of international communications and the increased complexity of the requirement. Thus, this book will deal as much with standards defining ISDN as it will with examples of ISDN in practice. Consequently, the book must, at some stage, address the question of whether this is the optimum way to develop the ISDN or, at least, explain why this way, certainly open to criticism, has been used.

1.5 Structure of the study

So far in this chapter we have assembled a series of pictures of what the ISDN intends, what it will give to the user, the objective requirements for its realization and the techniques necessary for its implementation. It is now appropriate to become more organized and draw up a structured plan indicating the format for the remainder of the book. Before this, however, the nature of the subject demands that the objective of the book be clearly expressed.

Hitherto, telecommunications specialists have been content to remain within their specialism. Switching was a specialism as was transmission. The advent of digital transmission and switching merged these and both switching and transmission experts have learnt a new, common specialism. A further widening of specialization has been caused by SPC, requiring the switching expert to practise also in system software applications. The advent of integration causes a veritable explosion in the technical area to be covered, because now the communications network is expected to serve a diverse variety of users, both persons and machines. The success with which the network will serve these diverse applications will depend in large measure on the clarity and suitability of the standards and the degree to which the various technical specializations co-operate in establishing a coherent integrated network. The book, therefore, is intended to meet five objectives. These are to:

(a) introduce integrated communications to this much wider audience, including those expert in the user machine and service specializations;

(b) influence the later stages of the standardization process;

(c) assist with the early stages of system design for implementation;

(d) introduce the subject to those now using the ISDN or considering its use; and

(e) introduce the modern development of the ISDN into the broadband ISDN (B-ISDN).

To fulfil these aims this book will visit and revisit all the subject areas outlined in this first chapter in an order thought suitable to build a comprehensive picture of the ISDN as it is emerging, as it ought to develop and as it can be used.

To begin with, it is necessary to look at the current networks, the worldwide telephone network, of course, and also the other widespread networks, telex, packet-switched and circuit-switched data. Chapter 2 will perform this review, and also review existing services, both those using a dedicated network and those that make use of existing networks, e.g. viewdata, teletext, facsimile. The available ISDN tools and techniques will then be examined in Chapter 3, which considers terminals (now becoming increasingly intelligent) and also the protocols necessary to access these terminals over the local network. Chapter 4 will consider OSI as a tool and discuss also the CCS systems with an ISDN application; these include #6 and #7, as well as (in the UK) DASS 2 and DPNSS and their equivalents in other countries. Chapters 5, 6 and 7 perform the same function on behalf of the B-ISDN, introducing the essential B-ISDN tools: synchronous digital hierarchy (SDH) and asynchronous transfer mode (ATM).

The treatment thus far will have provided sufficient background to introduce current operational and trial systems, and Chapter 8 will cover these briefly. It has been said that standardization is an essential tool for the ISDN, and this will also be considered in Chapter 8. The standards are emerging from the ITU-T, but the generating imperatives come from diverse sources in individual countries (notably the USA) and from international groupings (EU, NATO, the old Warsaw Pact countries, etc.). The nature of these imperatives and their influence will be outlined. The structure of the forums in which the standards are defined will also be described.

This more political section will be followed in Chapters 9 and 10 by a return to the technology surrounding the search for solutions to the problems impeding the ISDN and the B-ISDN. Such problems include subscriber loop signalling and error control, but also less immediately technical areas such as the allocation of costs and charges between services and networks and the various ways in which the ISDN can be made more attractive to the user. Finally, the account is completed by returning to the user and describing in detail the ISDN as it appears (or will appear) to the user (person or machine). What changes in user behaviour will the ISDN demand and encourage? How can the user influence the development of the ISDN and how can the user's requirements best be served by the ISDN?

All this treatment builds towards a consideration of the ISDN itself. The network (or networks) is considered fully in the final sections of Chapter 10. What network is implied by the standards as they exist today? What network would be ideal and what network is practically possible and should form the goal of continued standards-setting activity?

This final chapter attempts to survey the subject and draw conclusions in the light of the knowledge and understanding that has been provided by the book as a whole.

Chapter summary

The concept of the ISDN has been introduced initially in subjective terms and the scale of the endeavour has been illustrated by a review of all the existing and proposed component services and networks. This has enabled the proposed advantages of integration to be summarized. A more objective summary of the ISDN requirements was then reviewed as these impinge on existing local networks and terminals and, more generally, on the public network.

The review then moved on to the tools available and necessary to the introduction of the ISDN – SPC, CCS, LSI and standards being the techniques and methods essential. Finally, the objective and plan of the book were outlined, emphasizing that this is a tutorial book written to satisfy a real need for knowledge and clarification by the much wider audience requiring instruction in communications technology because of the ISDN.

CHAPTER 2

The network before ISDN

2.1 Introduction

To understand the need for the ISDN and the requirements for its implementation we must first understand the established services and the nature and requirements of the existing networks that are to be converted to the ISDN. This preparatory treatment must necessarily be brief and must assume that the reader has considerable prior understanding of telecommunications. As background reading, references 2, 9 and 10 are suggested.

The proposal to introduce the ISDN is predicated upon the existence of a digital network and the possibility of providing the digital interface at the user's premises. The worldwide public telephone network, however, is presently predominantly analogue, and much of it is likely to remain so well into the next century. Most network providers have, however, already at least made the decision to change to a digital network. Many have already commenced implementation of this change. Since 1990, all orders placed for new and replacement network equipment have been for digital equipment. The change to digital networks will occur at varying rates. As an example, the US urban network has been equipped with modern SPC analogue switches only relatively recently, and elements of the network will not be changed to digital until these switches have served their economic life. Another example is the former USSR, which at the time of its dissolution had not announced any intention of installing digital network elements. The digital network that the ISDN will use initially will appear in some places as a thin overlay on a predominantly analogue network with remotely located simplified exchanges, concentrators and multiplexers extending access to the digital overlay to the subscriber. To understand the implications and needs of ISDN we must understand the existing predominantly analogue telephone network, the emerging digital communications network and the existing and proposed dedicated services networks.

2.2 The current analogue network

Figure 2.1 attempts to encapsulate the significant features of the public switched telephone network as it is at present. Like all such summary statements, it cannot by any means represent the whole story. The reader must refer to the literature to obtain a comprehensive picture. For our present purposes the important feature of Figure 2.1 is the variety, often the unavoidable variety, of technologies that are used. These technologies are summarized in the figure as transmission, signalling and control, and they will be dealt with under these headings.

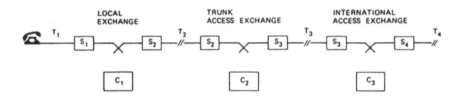

T_1 Local line plant 0.04 mm, 2-wire line.
S_1 Loop-disconnect or multi-frequency push-button signalling.
C_1 Direct control, register control or SPC.
T_2 Two-wire physical, local carrier, PCM over wire lines or co-axial cable, or microwave radio in rural areas.
S_2 Loop disconnect, long distance DC, AC signalling, inter-register MF signalling (typically R2)
C_2 SPC
T_3 Four-wire carrier or PCM systems or co-axial cable, microwave radio, etc.
S_3 Inter-register MF signalling (typically national variant of R2).
C_3 SPC
T_4 Four-wire carrier or PCM via cable or satellite.
S_4 Inter-register MF signalling (CCITT R2 or No.5) or CCS (#6 or #7)

Figure 2.1   The present analogue telephone network.

2.2.1 Transmission

2.2.1.1 In the local network (T_1)

Local network line plant ranges from multipair cable in underground ducts in urban areas through duct cable plus overhead distribution in suburban environments to overhead cable using party-line, pair gain and carrier telephone systems in rural areas. In the urban and suburban areas the cable network represents a significant investment and must be utilized by new technology as it is too expensive to throw away. It also has a long economic life, which is another reason for its continued use. Open wire lines and carrier telephone systems in rural environments do not represent a major investment but are expensive to maintain. It is therefore advantageous to replace these with new technology if this has not already been done. Optical fibre cable may stimulate further evolution, particularly for high-capacity business applications. Fibre in the local network will be used, sometimes in conjunction with radio, for the final fraction of a kilometre to the subscriber and sometimes in conjunction with cable TV or common antenna TV (CATV). Indeed, ISDN, and to an even greater extent B-ISDN, will encourage the merging of the local telephone network with CATV networks.

2.2.1.2 In the local exchange network (T_2)

Similar comments apply to the local exchange network with some additions. The high cost of laying new ducts in urban areas has already led to the frequent use of pulse code modulation (PCM) and analogue multiplex techniques to utilize the existing cable network more effectively. Many of the PCM systems used in this application in Europe are suitable for ISDN use. In the USA, however, the T1 system has been in use for many years, largely in the less densely populated environments, and is not compatible with the ISDN requirements. Where PCM or carrier systems are in use the local exchange network will be four-wire, but on physical circuits it will be two-wire.

In rural areas of less developed countries microwave radio links between exchanges are widespread.

2.2.1.3 In the trunk network (T_3)

Within the trunk network CCITT requirements for overall transmission quality imply that the network must be four-wire, near-zero loss. Transmission media are coaxial cable, microwave radio or, rarely, carrier systems on pair cable. In continental networks, such as in North America or Russia, trunk circuits are routed via satellite systems. Transmission systems are analogue carrier or PCM, although the penetration of the latter may not be as great as in the local exchange network where the economics of PCM have been more favourable. PCM came later to trunk systems because frequency division multiplexes (FDMs) were already widely deployed and provided the economies that PCM brought to local exchange networks.

2.2.1.4 In the international network (T_4)

In the international network, carrier systems over coaxial, fibre or satellite circuits predominate. Short-wave radio channels are rarely used nowadays. International microwave radio links exist throughout Africa, Asia, Eastern Europe and South America. Long-distance international traffic uses a global submarine cable network that enables a judicious mixture of satellite and cable routings, preventing the undue delays caused by several satellite "hops".

2.2.2 Signalling

2.2.2.1 In the local network (S_1)

Signalling in the local network is still predominantly loop disconnect. Many subscribers have telephones equipped with pushbuttons, but often conversion to dial pulsing takes place inside the telephone. Multifrequency pushbutton signalling (MF4) is widespread in the US and France, where networks were upgraded early with SPC exchanges, and it is now penetrating most networks as all exchange systems available are SPC and equipped for MF customer signalling. There used to be an economic disadvantage to moving away from loop disconnect because of the need to replace every telephone. However, in countries with liberated networks this is no longer the case as most subscribers have been persuaded by marketing campaigns[1] to change their telephone at their own expense.

2.2.2.2 In the local exchange network (S_2)

Signalling in the local exchange network depends upon the vintage of the exchange equipment and also quite markedly on the "style" of the administration. Direct control systems encourage loop disconnect or other DC signalling systems or their AC equivalents. Register control systems use increasingly sophisticated systems, a modern example of which is multifrequency inter-register signalling, such as CCITT R2. In the UK, administration "style" has necessitated the sending of meter pulses down from the trunk access exchange to the subscriber's meter in the local exchange. "Style" has necessitated an esoteric MF signalling system in France and in the US relatively unsophisticated outband AC signalling. All such signalling is channel associated: the signals are passed over a bearer circuit (channel, time-slot, physical pair) associated with the eventual conversation path.

2.2.2.3 In the trunk network (S_3)

Most of the worldwide trunk network is now equipped with multifrequency inter-register signalling, most probably R2, of which there are many national variants. (The UK variant is known as SSMF 6 and includes a major extension into new signal groups but, after all the fuss of defining a "UK-only" system, it was never used.) The US system has, for more than 10 years, had a national CCS network based initially on CCITT #6 but now mainly based on digital CCS, CCITT #7, which is being used increasingly frequently throughout the world.

---

1. Sadly, in the UK, the marketing campaign has preceded the application of the technology so that many subscribers have purchased pushbutton telephones using loop disconnect, which must be changed again as MF4 signalling becomes available.

2.2.2.4 In the international network (S_4)

International signalling has to include additional functions; transit signals, language digits, the disabling of time assignment speech interpolation (TASI) during signalling and testing are examples. The international versions of R1[4] and R2[5] are in common use as are CCITT signalling systems 4 (two-frequency) and 5 (multifrequency)[8]. All these systems are inter-register signalling systems. CCITT #6 CCS was defined for international use but has achieved little penetration outside the US. CCITT #7 is defined as the digital system mandatory for international use and is being used increasingly as digital international gateway exchanges are brought into service.

2.2.3 Control

2.2.3.1 In the local network (C_1)

Direct control (typically Strowger), register control of marker systems and SPC systems of varying sophistication are all in widespread use. SPC has massive penetration in the US, particularly in the old AT&T network with ESS 1, 2, 4 and 5. In the UK, TXE2 is notionally common control but with very limited features. TXE4 is SPC, but again with limited features. Later marks of TXE4 (TXE4A) are fully SPC and the earlier version has been upgraded to the same level.

The features supported by the local exchange, of whatever technology, are very dependent on administration "style". The UK, for example, uses meter pulses whereas the USA has used call charging for many years and has many operator interception features, for calls from hotels, for example.

2.2.3.2 In the trunk network (C_2)

Because the introduction of subscriber trunk dialling has, in many cases, coincided with the introduction of SPC, trunk access and trunk transit exchanges are almost invariably SPC. Apart from some features of the access exchange (charging, routing) the requirement for trunk exchanges is much simpler than that of the local exchange.

2.2.3.3 In the international network (C_3)

It is the control of international switching centres that may have much to teach the designers of switching control machines to be used in the ISDN. International subscriber dialling has been introduced only during the past 30 years, and control of international switching centres is universally SPC. Previous switching centres were at best semiautomatic. Such centres provided for the involvement of an operator at the originating gateway only, subsequent call treatment being automatic. Even in semiautomatic working the international aspects introduce requirements new to telephone switching. Chief among these is the need to provide for bulk accounting of the different traffic flows in order to apportion the cost of the call between several co-operating administrations. This problem is analogous to, but less detailed than, the need to apportion costs of a connection over the ISDN between the various network operators and service providers.

2.3 The emerging digital network

Figure 2.2 depicts, in the same encapsulated format, the digital network that is currently appearing, usually as a thin overlay of the existing analogue network. It can be seen at once that the variety of transmission, switching and control applications is substantially diminished and that the sophistication of control and signalling, present only at the higher levels of the analogue network, has pervaded the entire digital network.

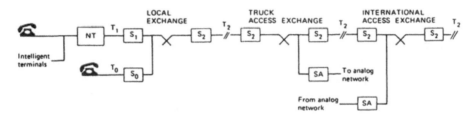

S_0/T_0 Digital to analogue conversion, multi-frequency push-button signalling
NT ISDN user terminal.
T_1/S_1 30/24 channel PCM local system with local CCS.
T_2/S_2 30/24 channel PCM and higher order multiplexes. CAS or CCS.
C SPC in all cases.
S_A Analogue signalling CAS (see Figure 2.1)

Figure 2.2 Future digital communications network.

2.3.1 Transmission

2.3.1.1 In the local network (T_0, T_1)

Some of this has been oversimplified as the variety of ISDN access has not been shown to avoid pre-empting later treatment. Both T_0 and T_1 will use the same physical bearers as exist in the existing analogue network. New installation of local cable plant could be very different using optical fibre or sharing with local cable TV distribution or using locally sited multiplexers and concentrator exchanges with conventional two-wire local end lines or using radio for the final few hundred metres. The 30- or 24-channel digital multiplex, shown as T_1, connects subscribers with telephone instruments that perform the digital-to-analogue conversion by means of some form of multidrop and connects digital PABX equipment with, or more probably without, multidrop.

26

2.3.1.2 In the remaining network (T_2)

Throughout the remaining network the transmission media must be suitable for PCM at 2.048 Mbit/s or higher order PCM multiplexes. Such media are four-wire line without loading coils, coaxial cable, microwave radio, optical fibre, etc. Note that the use of PCM over microwave radio or coaxial cable provides a reduced channel capacity when compared with FDM techniques.

2.3.2 Signalling

2.3.2.1 In the local networks (S_0, S_1)

For analogue telephones the signalling is identical to that required by the analogue network, which, for new installations, is predominantly MF pushbutton signalling. Digital local exchanges are still specified to accept loop disconnect signalling on existing lines. For digital telephones (probably ISDN subscribers) or digital PABXs, the signalling will be the defined ISDN signalling system. Channel 16 is available for channel-associated signalling, but CCITT requirements for digital subscriber signalling system no. 1 (DSS 1)[11] and all existing or proposed ISDN implementations define a CCS approach. The UK defined a system known as digital access signalling system (DASS 2) in advance of full CCITT definition.

2.3.2.2 In the remaining digital network (S_2)

Throughout the remainder of the network the signalling system is, and will be, CCITT #7. The US is an exception to this; there CCITT #6, adjusted slightly to US requirements, is already in use in the digital trunk network (see Chapter 7 of reference 2).

2.3.2.3 At the interfaces to the analogue network (S_A)

The emerging digital network and the existing analogue network must, of necessity, interconnect. Most plans for digital introduction envisage a digital overlay technique using strategies to limit the complication and expense of analogue-to-digital conversion and the necessary signal conversion required at the interfaces (Fig. 2.3).

The choice of interface position in the network will be such as to limit the amount of signal conversion and to encourage connections entering the digital overlay to remain digital for as long as possible. Even with such careful planning the interface signalling adaptation will be complex, expensive and, of course, eventually redundant.

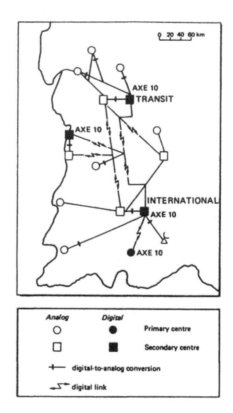

Figure 2.3 Digital overlay[12].

2.3.3 Control

The digital network demands SPC, and this exists throughout the network. CCS provides facilities for control at each node to co-operate with other nodes so that complex functions can be provided as a service to the complete network.

2.4 The telex network

Figure 2.4 shows, in similar format to Figures 2.1 and 2.2, the telex network. The UK telex network development is described in reference 13. As a continuation of the telegraph techniques from which they emerged, telex machines used 80V DC signalling. International teletypewriter alphabet no. 1 (ITA 1), similar to ASCII code (see §2.6.1), was developed to encode the alphanumeric keyboard characters. Again, because of telegraph practice, the machines were designed to give an automatic "answer back" confirming the called party identity. In contrast to a telephone call therefore, a telex call is

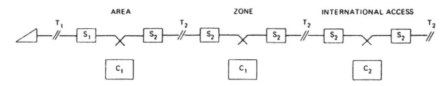

T$_1$ Local line plant.
S$_1$ 80 V or VF signalling.
C$_1$ Direct control, register control or SPC.
T$_2$ Similar to analogue telephony network.
S$_2$ CCITT defined systems Type A, B, C, or D.
C$_2$ SPC

Figure 2.4   The existing telex network.

confirmed and the called station need not be attended. There is, however, another called party unavailable possibility if the power to the called machine is not switched on.

Using direct control or simple register control of switching it was usual to provide selection signals from a dial. Introduction of SPC allowed the use of keyboard selection, and this is universal on modern systems. SPC switching also allows such features as store and forward (when the called machine is switched on), multiple addressing and call redirection With keyboard signalling there is much more uniformity across the network with less variety of signalling techniques. Modern networks and machines use voice frequency (VF) signalling, which is more compatible with solid state components.

Because of its legal status and worldwide network the telex service will persist for some time and is already interfaced by the more modern text communications services and networks such as packet switching and teletex.

2.5 Packet-switching networks

The packet-switching service (PSS) is so new that it has been subject to CCITT recommendations almost from inception and has not spawned a multiplicity of interfacing protocols. There are, nevertheless, some 60 separate PSS networks in the world and these can exchange messages through universal use of the CCITT X.75[14] protocol between networks. Typical applicable protocols within the networks are shown in Figure 2.5, but a variety of protocols exist in the different networks.

Although the PSS is devoted to data only, the size of the packets and the frequency with which packets of the same message are transmitted does not much matter. Provided packet size is kept small and frequency of sending is

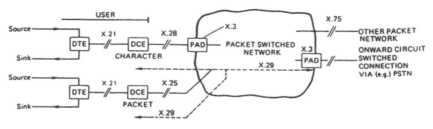

DCE	Data circuit terminating equipment
DTE	Data terminal equipment.
PAD	Packet assembler/disassembler. CCITT Recommendation X.3.
X.21	Interface between DTE and DCE.
X.25	Interface between DTE and DCE and public network.
X.28	Access from character DTE to PAD.
X.29	Exchange between PAD and packet DTE or another PAD.
X.75	Data transfer between different packet switching networks.
X.3	PAD definition.

Figure 2.5 The existing packet-switched network.

kept high there is no reason why voice traffic should not pass over a suitably engineered PSS network; this is the concept underlying asynchronous transfer mode (ATM).

The protocols developed for the PSS have a relevance to and have had an influence upon the development of protocols for the ISDN. As such, they will be treated in more depth later in this chapter. The principle of packet switching has been illustrated already in Figure 1.4.

The switch required in a packet-switching exchange is shown, in concept, in Figure 2.6. A variety of ports are shown: X.25[15] ports, X.75 ports and an internal network port.

Packets arriving at a port are buffered pending service by the common highway. The highway protocol conveys the packet to a designated processor, where it is processed by a virtual circuit handler task. If the packet is a call control packet then the processor may have to refer to the record of terminal and network characteristics contained in the common memory to determine initial routing and ascertain that the desired routing is possible (i.e. the terminals are compatible). For data packets, information on the virtual circuit routing will already exist in the processor's local memory.

The virtual circuit handler task determines the outgoing port to which the packet is to be routed, and the packet is then transferred via the highway protocol to the designated port. There the packet is queued in the output buffer until it is serviced by the connecting link or trunk.

Commonly, the virtual circuit handler task deals with the packet level of X.25 or X.75 and the port handler deals with the link level. This is illustrated for the message passing through the system in Figure 2.6.

This description has used the term "virtual circuit" to describe a method of

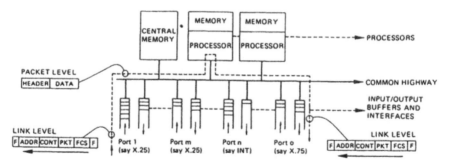

Figure 2.6 Concept of the packet-switching exchange[16].

message passing in which a circuit is not dedicated to the connection but in which the service perceived by the user is indistinguishable from dedicated circuit working.

2.6 Data networks

The development routes adopted for data networks have had a profound effect on the way that the concept of the ISDN has been expressed. Thus, in the remainder of this chapter, data networks and their codes and protocols are treated in considerably more detail than in the preceding sections. It must be emphasized that the packet-switching network is a type of data network and can be thought of as a goal towards which data network development has tended. This development pattern may well have been upset by the emergence of the ISDN. A more likely scenario is that packet-switching and integrated services networks will eventually become indistinguishable.

A personal view of the development of data networks is provided by Figure 2.7. Tracing the story provides a suitable introduction to the concept of protocols, which has become central to the development of integrated networks.

2.6.1 Star network

The development of the computer led to the need for a man–machine interface. A suitable interface is the typewriter keyboard, already used in telex working, and this became the input medium for man–machine communication (MMC). Communication from machine to person was provided first by teleprinter printout and later by the addition of the visual display unit (VDU). However, communication from keyboard to computer or from computer to printer or VDU required a means of converting the 130 or so different characters into digital code. One such code with still nearly universal application is

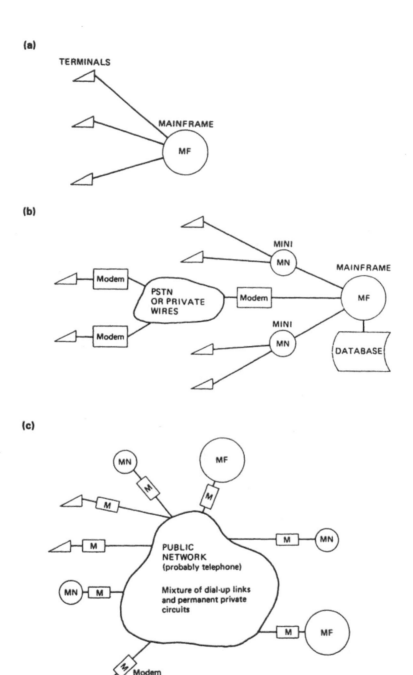

Figure 2.7 Development of data networks. (a) Star network of "dumb" terminals. (b) Minicomputers, mainframe and remote access. (c) Dial-up and dedicated data network using public network.

32

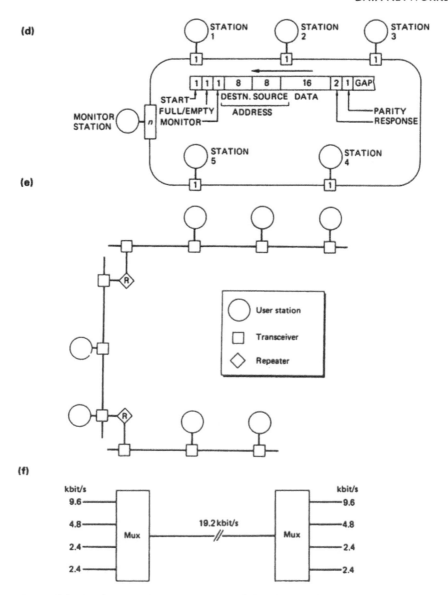

Figure 2.7 Development of data networks. (d) Local area network: Cambridge ring. (e) Local area network: Ethernet (reproduced with permission from ref. 18). (f) Networking with multiplexers.

the ASCII code (United States of America Standard Code for Information Interchange) (Table 2.1).

Figure 2.7a shows the earliest "network", in which several users are allowed to use the computer at once, sharing the computer resource. To achieve this the computer must perform some kind of time-division switching

33

(g)

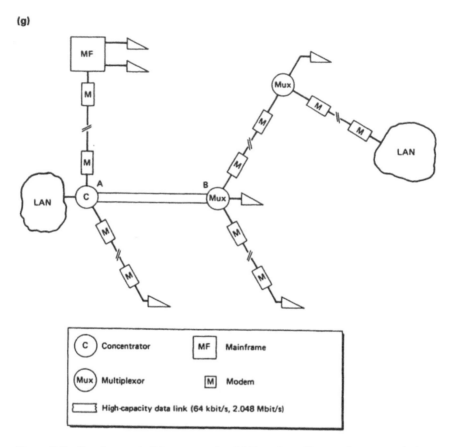

Figure 2.7 Development of data networks. (g) Complex wide area data network. (reproduced with permission from ref. 18).

to share input and output facilities between the several users. Furthermore, the connecting cable between the terminal and the computer must be defined. This terminal interface was first defined in its present-day form by the US Electronic Industries Association (EIA) as EIA standard RS232. This standard defines not only the electrical nature of the interface but also the mechanical details of the plug and socket. The electrical interface only is defined again in CCITT Recommendation V.24, so that the interface is known interchangeably as RS232 or V.24[17]. The interface is illustrated in Figure 2.8 conveniently introducing the abbreviations DTE and DCE.

2.6.2 Remote access

Increased component integration and increased speed of operation led, in time, to much more powerful mainframe computers and to the emergence of

Table 2.1 ASCII codes.

$b_3b_2b_1b_0$	Row (hex)	$b_6b_5b_4$ (column)							
		000 0	001 1	010 2	011 3	100 4	101 5	110 6	1 1 1 7
0000	0	NUL	DLE	SP	0	@	P		p
0001	1	SOH	DC1	!	1	A	Q	a	q
0010	2	STX	DC2	"	2	B	R	b	r
0011	3	ETX	DC3	#	3	C	S	c	s
0100	4	EOT	DC4	$	4	D	T	d	t
0101	5	ENQ	NAK	%	5	E	U	e	u
0110	6	ACK	SYN	&	6	F	V	f	v
0111	7	BEL	ETB	'	7	G	W	g	w
1000	8	BS	CAN	(	8	H	X	h	x
1001	9	HT	EM	)	9	I	Y	i	y
1010	A	LF	SUB	*	:	J	Z	j	z
1011	B	VT	ESC	+	;	K	[	k	{
1100	C	FF	FS	,	<	L	\	l	|
1101	D	CR	GS	-	=	M	]	m	}
1110	E	SO	RS	.	>	N	^	n	~
1111	F	SI	US	/	?	O	_	o	DEL

Control codes

NUL	Null	DLE	Data link escape
SOH	Start of heading	DC1	Device control 1
STX	Start of text	DC2	Device control 2
ETX	End of text	DC3	Device control 3
EOT	End of transmission	DC4	Device control 4
ENQ	Enquiry	NAK	Negative acknowledge
ACK	Acknowledge	SYN	Synchronize
BEL	Bell	ETB	End transmitted block
BS	Backspace	CAN	Cancel
HT	Horizontal tab	EM	End of medium
LF	Line feed	SUB	Substitute
VT	Vertical tab	ESC	Escape
FF	Form feed	FS	File separator
CR	Carriage return	GS	Group separator
SO	Shift out	RS	Record separator
SI	Shift in	US	Unit separator
SP	Space	DEL	Delete or rubout

minicomputers and, later, the personal computer. Networks could now consist of numerous users accessing a computer bureau based on a very powerful mainframe or numerous users accessing numerous personal (micro-), mini- and mainframe computers. Both these arrangements are illustrated in Figure 2.7b. To achieve such networking it is necessary to convert the digital data streams into a form suitable for passing over analogue transmission lines within, in general, the speech band of 4 kHz. The device that achieves this is a modem (*mo*dulator/*dem*odulator), which converts digital ones (1s) and zeros (0s) into pulses of carrier frequency.

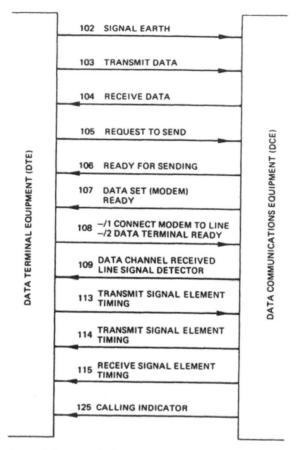

Figure 2.8 CCITT V.24 interface.

Clearly, sending and receiving data will only be successful if the terminal devices receive information in a structured form that they can recognize and process. Networks such as those in Figure 2.7b, c and e are therefore only viable if all the terminals can adopt the same protocol. Industry standards were soon developed by individual manufacturers (e.g. CO1, CO2 and CO3, DEC VT 100, IBM SNA).

The first nationwide networks or wide area networks (WANs) had relatively limited terminal functions. Early examples were airline booking systems and networks for bank teller machines. With such limited, if not simple, functions, fairly simple methods of polling the terminals could be adopted.

2.6.3 Local area networks (LANs)

As the power of the mini- and microcomputers increased there emerged a need for very powerful conglomerations of computer facilities within a restricted area, such as a university campus or an industrial research facility. This introduced solutions based on LANs, and two examples are depicted in Figures 2.7d and e.

The Cambridge ring (Fig. 2.7d), although now of only historical interest, is a good illustration of another type of protocol. A terminal wishing to transmit scans the circulating data word looking for "empty" in the second bit. On finding "empty" (0), and if the station has a message to send, it changes it to "full" (1) and inserts address and data as appropriate. A station seeing the full/empty bit marked "full" (1) can do one of three things.

(a) If it filled the slot, it writes 0 in the full/empty bit and 0 in the adjacent monitor bit. The slot is thus passed on marked "empty".

(b) If it did not fill the slot, it checks the destination address and, if it is this station address, reads source address and data and updates response bits as necessary.

(c) If it did not fill the slot and the destination address is for another station, it passes on the slot without further action.

The monitor station is watching the monitor slot to ensure that messages are being received. A monitor "0" is changed to "1" as it passes the monitor station and, as a sending station rewrites monitor to "0" when it sees its message return, the monitor station will know that a "full" slot with "1" in the monitor bit indicates a message that is eternally circulating around the ring. The monitor station must then reinitialize the slot pattern and record a system failure.

The circulating slot word is 38 bits long and must be separated from preceding and succeeding words by at least one empty bit (set to "0", zero); in practice a gap of two empty bits was used. The ring must, therefore, have a total delay of 40 bits or multiples of 40 bits. Usually, however, a single word circulating around the ring is adopted. Each station introduces a 1-bit delay and the remainder is provided by a shift register at the monitor station.

The Cambridge ring is an example of an *empty slot* ring strategy. An alternative strategy is *token passing*, in which the token code, detected by a station wishing to send, is extracted from the circulating word and reinserted at the end of the message. Ring LANs are termed active networks when every station is involved in the message passing. Ethernet (Fig. 2.7e) is an example of a bus configuration LAN in which only the transmitting and receiving stations are active. Such networks are termed passive networks.

LANs typically operate at a data rate of 10 Mbit/s. There is a need, therefore, for transmission media for such networks to be high capacity. The office automation revolution had a significant impact on our working environment, at least in terms of the quantities of coaxial cable that had to be installed. It is thus a relief to be offered, in the ISDN, a telephone network providing

144kbit/s to each user terminal over ordinary wire pairs, obviating the need for special data networks to the user.

2.6.3.1 Fibre-distributed data interface (FDDI)

Ethernet, for example, operates at a speed of 10kbit/s but, because of overheads and delays, its information transfer rate in practice is considerably less than this. If an Ethernet LAN grows to a size of more than about 80 devices then special design methods must be used to provide users with an acceptable speed of operation. If, working on your stand-alone PC, it is possible to access files almost instantaneously, users will not be best pleased if, on the LAN, they must wait 30 seconds to two minutes for a file.

Increasingly, also, it is becoming economical to wire up the workplace with optical fibre. There is a need for a LAN protocol that will use the much greater bandwidth available with fibre.

FDDI operates at 100Mbit/s, and more recent developments upgrade this speed to 200Mbit/s. It is a ring system but with much greater security than the low-speed LANs. Information is transmitted by all devices at 100Mbit/s in both directions around the dual, contra-rotating ring. If any device or any link fails then the neighbouring devices loop back the information so that the remainder of the ring still operates (see Fig. 2.9).

FDDI is defined as a transport mechanism. It has the advantage that it will carry communications expressed in most existing protocols. This permits it to be used as the backbone LAN serving Ethernet or other low-speed LAN branches. The FDDI devices would be routers, bridges or gateways interconnecting the FDDI LAN with the departmental LANs running at lower speeds. Thus, like the Ethernet branches, information would be carried over the FDDI backbone using the Ethernet protocol. One candidate protocol often used is transaction control protocol (TCP). In much the same way as X.25, the efficiency of such protocols is reduced when the large majority of individual packets are not part of the same, or a few, continuous messages. Networking traffic experts describe this, packets belonging to the same, or a few, continuous messages, as the "common case" scenario. As packet sizes become larger and the number of different connections handled by the large-capacity transmission medium increases, the incidence of adjacent packages being different increases. Thus, the capability of nodes caching address information and using it to route subsequent packages quickly and efficiently is reduced. This topic will be considered again later when discussing the characteristics of ATM.

2.6.4 Multiplexers and concentrators

As data networks increased in size and served ever more sophisticated functions, it became economic to provide high-capacity long-haul data links and share them with many circuits by multiplexing, (Fig. 2.7f). A degree of semi-

(a) Ethernet or other subnetwork

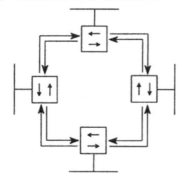

(b)

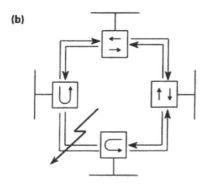

(c)

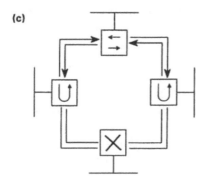

Figure 2.9 Fibre-distributed data inter-change local area network. (a) Normal operation. (b) Operation with a failed link. (c) Operation with a failed node.

permanent data switching was introduced by using concentrator switches to connect many little-used data links on to high-capacity high-usage links. Figure 2.7f illustrates conventional multiplexing, but an alternative form of multiplexing, statistical multiplexing, combines the functions of multiplexer and concentrator by sharing the single transmission resource between many users in proportion to the user activity.

2.6.5 Wide area networks (WANs)

All these techniques are illustrated as components in a wide area network in Figure 2.7g. The link between concentrator A and multiplexer B could be a 2.048 Mbit/s link that, instead of being used for a 30-channel CEPT link, is submultiplexed as necessary and devoted to data traffic. In the UK such a facility is marketed as Megastream for use only between customer premises; the provision of a single CEPT channel (64 kbit/s) for private data is marketed as Kilostream.

2.7 Codes and protocols

The tour through data networks just concluded has enabled us to gain a preliminary insight into the concept of and need for protocols. In telecommunications, we have always been conscious of the need for signalling codes, but only relatively recently has the term protocol been introduced, and the difference between the terms is not always clear. That this lack of clarity is not confined to the engineering profession is illustrated by Table 2.2, which contrasts two dictionary definitions. Table 2.3 offers as definition the author's understanding of the words when used in the context of telecommunications and data processing.

In the earlier parts of this chapter, examples of codes (ASCII) and protocols (the Cambridge ring protocol, packet-switching protocol, Fig. 2.6) were introduced, so readers should already be aware of these ideas and the need for them. The need can be illustrated a little better by considering communication between two people by letter (Fig. 2.10). The recipient will not act on the message unless it carries an authorized signature (password) confirming that it really comes from the sender. Nor is the recipient likely to act expeditiously unless the letter is couched in acceptably polite terms. A better example here might be the Oxford Dictionary definition 5 (Table 2.2). A papal bull, for example, only has validity if the preamble and conclusion use the correct formulae. In telecommunications these functions are all part of error detection, error correction and, perhaps, encryption. Address information is evidently fundamental and the completed missive must be packed in a convenient form for the postman. Figure 2.10 also shows how this whole process can be seen as a succession of levels of abstraction from the communications link (the postman), each level causing a new outer covering to be wrapped around the central message. Current ideas on universal protocols for machine-to-machine communication adopt exactly this approach.

The corresponding diagram showing, in greatly simplified form, how a message is transmitted across the ISDN is shown in Figure 2.11. Each descending layer in the hierarchy adds its own header and, perhaps, a trailer to the

Table 2.2 Code and protocol – the dictionary definitions.

	Webster	Oxford
Code	Derivation: Latin *caudex*: trunk of tree, tablet of wood covered with wax for writing	
	1. A systematic statement of a body of law	1. One of the systematic collections of statutes made by the later (Roman) emperors
	2. A system of principles or rules	2. A digest of laws of a country, or of those relating to any subject
	3. A system of signals for communications	3. A system of rules and regulations on any subject.
		4. A system of signals
		5. A collection of writings
		Addendum 4b
		Any system of symbols and rules for expressing data and instructions in a form usable by a computer or other equipment for processing or transmitting data (1946)
Protocol	Derivation: Greek *Protokollon* (prot + kollon, to glue to), first sheet of papyrus roll bearing date of manufacture	
	1. An original draft, minute or record of a document	1. The original note of an agreement duly attested, forming the basis for any subsequent deed.
	2a. A preliminary memorandum of diplomatic negotiation	2. The original draft, minute or record of a diplomatic document
	2b. The records or minutes of a diplomatic conference or congress	3. Formal statement of a transaction
	3. A code of diplomatic or military etiquette and precedence	4. In France. Etiquette to be observed by the head of state
		5. Official formulas used at the beginning and end of a charter, papal bull, etc., as distinct from the text
		Addendum 4b
		Rigid prescription or observance of precedence and deference to rank as in diplomatic and military services; official etiquette and formality (1945)

Table 2.3 Code and protocol – definition for telecommunications.

Code	Any system of symbols and rules for expressing data and instructions in a form understandable by a computer or other equipment for processing or transmitting data
Protocol	The formula for arranging the parts of a message consisting of coded data and instructions into a form usable and understandable by communicating data processing devices

1) Agree language understood by sender and recipient.	CODE	
2) Compose and write text.	PROTOCOL	MESSAGE
3) Employ polite formulae at beginning and end of message; e.g.: Dear Sir . . . Yours faithfully Dear Anne . . . Yours sincerely		CHECK Redundant information enabling validity of message to be assured.
4) Signature		PASSWORD SENDER IDENTITY
5) Address		ADDRESS
6) Copies to:		BROADCAST ADDRESSES
7) Envelope and correct stamp		TRANSMISSION INSTRUCTIONS

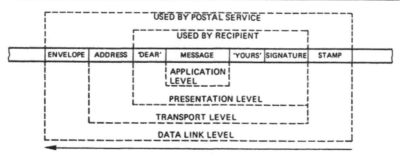

Figure 2.10 Communication by letter.

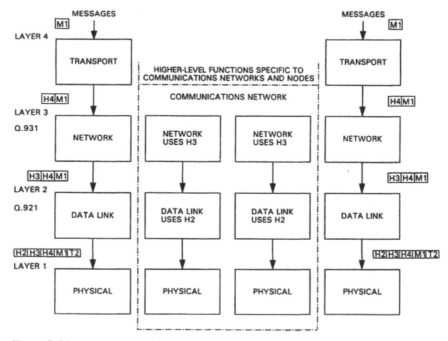

Figure 2.11 ISDN message hierarchy.

message, and intermediate nodes consult this header information in determining how to pass on the message.

In this description the "envelopes" are information directed at the postman, the communications devices, and have become known as the information for the control plane, the C-plane. The message itself is directed at the communicating users, which may be human users or machines; this communication is taking place on the user plane, the U-plane.

The ISDN makes this clear logical separation of signalling information and message information or data. The C-plane determines the routing of the message, although the message itself is routed via the U-plane. In the basic ISDN the two planes may not have a recognizably separate existence but the principles applied use this separation. Thus, a truer picture of the operation of the ISDN than that of Figure 2.11 is shown, in even simpler form, in Figure 2.12.

The C-plane information is used to establish communication between the user devices, and the communication itself may take any one of a number of forms, a permanent (voice or high-speed data) connection over a B channel, a packet-switched virtual connection over a B channel or over the same D channel that was used in the C-plane communication, and there may be others.

The remainder of this chapter will describe a number of protocol elements relevant to the ISDN and place them in the context of the levels of the OSI model. This will at least enable the reader to understand the abbreviations used frequently in the literature (HDLC, LAPB, V.24, etc.) and it is hoped that it will make clear the relative importance and place of the various protocol techniques within the ISDN.

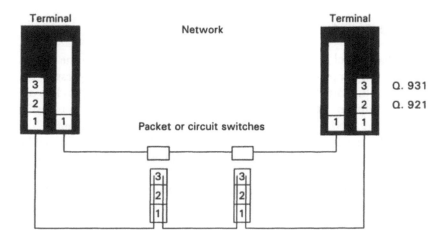

Figure 2.12 ISDN switched-mode services.

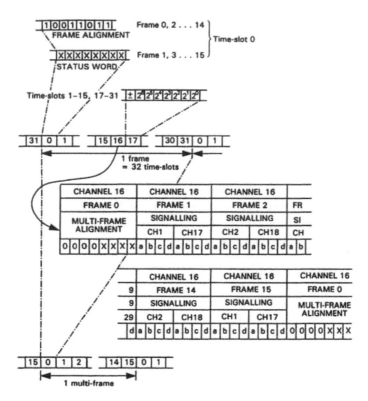

Figure 2.13 CEPT 30-channel PCM system[19].

2.7.1 Digital coding

At a very basic level it is assumed that the reader is familiar with the way in which speech is sampled and encoded into binary-coded words expressing the amplitude of the sample and how the binary-coded strings of 1s and 0s are modulated into an alternating waveform suitable for transmission using a method such as HDB3. (Chapters 3 and 4 of reference 2 and Chapter 7 of reference 18 are suggested reading for those in need of some revision.) The digital coding format will be dealt with further in Chapter 7 when discussing SDH, but Figure 2.13 illustrates the coding defined in ITU-T Recommendation G.732[19]. (Perhaps unfairly, this diagram was not included as a reminder in the first edition of this book.)

2.7.2 ASCII coding

For machines to communicate, a special "language" is required, and it has become accepted practice for machines to communicate using, at least, all the keys on a conventional (tele)typewriter keyboard. ASCII code (Table 2.1) has become a very common means of achieving this.

2.7.3 The interface

Having defined the communications code, it is necessary to define the transmission medium over which the coded message is to be sent. The medium itself may be one of many, and its designer should be free to develop it without exhaustive knowledge of the communicating people, machines and processes. It has therefore proved convenient to define interfaces to the transmission system, as shown in Figure 2.14.

The data terminal equipment (DTE) embodies everything that the data processing machine designer needs to know about the transmission system. Similarly, the data communications equipment (DCE) embodies all that the transmission engineer needs to know about the terminal devices. Between DTE and DCE lies an interface. Two examples are shown in Figure 2.14. In Figure 2.14b the interface is the two-wire telephone line that carries speech as an alternating signal, initiates seize with a loop and recognizes calling by a loop. Radio equipment cannot transmit the DC loop condition, necessitating a different interface that uses some form of tone signalling initiated by signals on the E lead and, when detected, repeated on the M lead (E & M signalling). Figure 2.14c shows the V.24 interface for data communications that we have seen already in Figure 2.8.

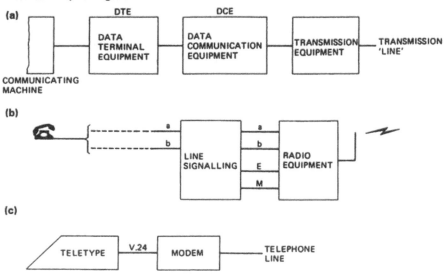

Figure 2.14 Fundamental interface concept. (a) General functional division. (b) Example: rural radio telephone. (c) Example: data modem.

2.7.4 Physical protocol

A glance at Figure 2.8 indicates that the V.24 interface defines all the functions required to set up a transmission so that streams of data can be sent over circuit 103 and received over circuit 104. It is not known where the data goes to or comes from at the far end of the link or how to interpret the data received or how to format the data sent. Thus, one end of the link expecting to receive ASCII-coded data would make little sense of digitally encoded speech.

2.7.4.1 Link protocol – link use

To make the data meaningful, we must give it a structure, i.e. show where it starts and where it ends. If the transmission medium introduces errors, as it invariably does, we must also introduce some form of error detection and correction.

One method of performing these tasks is known as high-level data link control (HDLC). Structure is provided by adopting a frame format (Fig. 2.15). A flag indicates the start of the message; the address bits will define sender and recipient; the control bits will indicate the nature of the message and provide facilities for messages to be acknowledged; and the check bits following the message will perform a check on the validity of the complete message excluding the flag.

The format of the control field indicates one of three types of frame; further categories within these main divisions are indicated by supervisory type bits and unnumbered frame modifier bits. Thus, just as the HDLC protocol has been used as an envelope to contain the information, it also has the capacity to perform a variety of functions.

This is a protocol allowing recognizable communications over a link with built-in error control mechanisms. These mechanisms are provided by the check sequence and also by the count mechanism, which allows messages to be acknowledged by updating the counts. However, there is still no control over when a terminal "speaks" and when it "listens".

2.7.4.2 Link protocol – link access

The control of the conversation, when to listen and when to send, is provided by an access protocol. Two such protocols are defined for use with HDLC: link access procedure (LAP) and link access procedure balanced (LAPB). Figure 2.16 gives their definitions and illustrates their effect. LAPB is becoming the most widely used as either the DTE or the DCE can initialize the link whereas in LAP both have to co-operate in initialization. Figure 2.17 illustrates a LAPB communication in HDLC between DTE and DCE and shows the format of the various messages.

A code to be used has been defined, as has a physical link, a protocol for using the link and a protocol for configuring the terminals to use the link (link access). Little has been said about the actual information to be transmit-

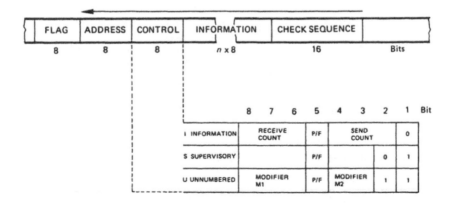

Figure 2.15 HDLC frame structure.

ted and nothing about how to ensure that the terminal users can interpret the information. As an example to illustrate these aspects we will return to packet switching.

2.7.4.3 The information packet

The information packet defined by CCITT Recommendation X.25 is illustrated in Figure 2.18 and Table 2.4. The three header octets of the packet contain sufficient information to identify and define both sender and receiver (after an initial exchange that includes called and calling addresses) and allow sequence counts of send and receive messages, thus providing the basis for message acknowledgement.

2.7.5 Review

This discussion began with the analogy of a posted letter (Fig. 2.10) and moved on to describe a series of codes and protocols that closely resemble the levels of abstraction illustrated in Figure 2.10. Figure 2.11 illustrated these levels more accurately, introducing a layer above the three layers for which

(a)

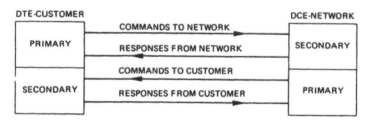

(b)

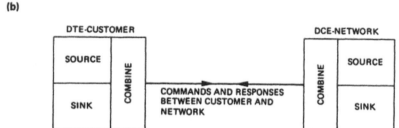

Figure 2.16 Link access procedures. (a) Link access procedure (LAP) configuration. This procedure configures the connection with two independent directions of transmission. (b) LAP balanced (LAPB). This procedure configures the link with a single full duplex transmission channel.

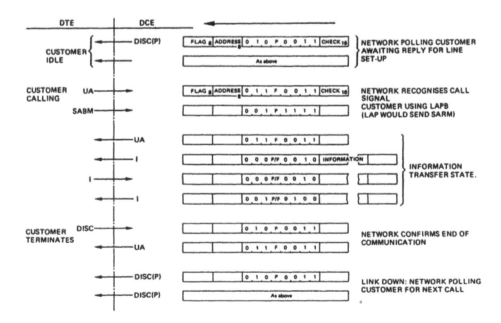

Figure 2.17 Call progress using LAPB.

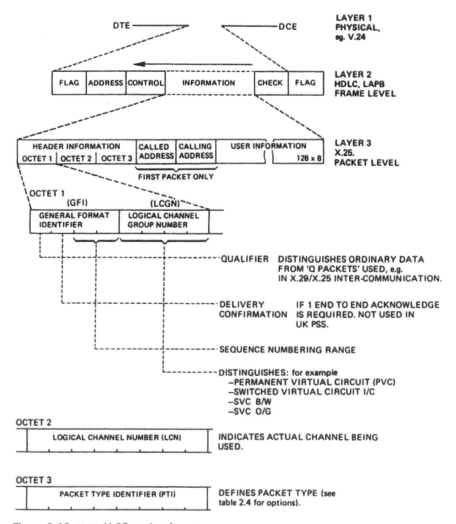

Figure 2.18 CCITT X.25 packet format.

example codes and protocols were considered. Table 2.5 recapitulates what has been covered.

The attentive reader will have noted the similarity between Figure 2.11 and the OSI model shown in Figure 1.7. Figure 2.11 is in fact a detailed representation of the first four layers of one implementation of the OSI seven-layer model. Figure 2.11 also shows that significant information is available in layer 3, the network layer, to effect switching and transmission of the message across the network. The switching machines involved, in general, need only concern themselves with the lower three layers.

As a final illustration of the universal nature of the concept of layers of

Table 2.4 Packet type identifier options

Packet type		Coding of bits in octet 3							
DTE → DCE	DTE ← DCE	8	7	6	5	4	3	2	1
Call request	Incoming call	0	0	0	0	1	0	1	1
Call connected	Call accepted	0	0	0	0	1	1	1	1
Clear request	Clear indication	0	0	0	1	0	0	1	1
DTE clear confirmation	DCE clear confirmation								
DTE data	DCE data	P(R)		M	P(S)	0			
DTE interrupt	DCE interrupt	0	0	1	0	0	0	1	1
DTE interrupt confirmation	DCE interrupt confirmation	0	0	1	0	0	1	1	1
DTE RR	DCE RR	P(R)		0	0	0	0	1	
DTE RNR	DCE RNR	P(R)		0	0	1	0	1	
DTE REJ		P(R)		0	1	0	0	1	
Reset request	Reset indication	0	0	0	1	1	0	1	1
DTE reset confirmation	DCE reset confirmation	0	0	0	1	1	1	1	1
Restart request	Restart indication	1	1	1	1	1	0	1	1
DTE restart confirmation	DCE restart confirmation	1	1	1	1	1	1	1	1

abstraction to digital communications, Figure 2.19, reproduced from reference 2, shows a similar layered concept as used in CCITT #7 CCS.

Chapter summary

Introductory passages outlined the nature of the existing communications networks, the present analogue telephone network, the emerging digital telephone network, the existing telex network and the packet-switched network. The discussion then looked back at the development history of data networks and, in relation to LANs, introduced other forms of protocol to those already considered briefly in relation to packet switching.

The discussion so far illustrated the importance of standardized codes and protocols to any form of machine-to-machine communication, and the

Table 2.5 Review of protocol

Layer 1 physical	V.24
Layer 2 data link	HDLC with LAPB
Layer 3 network	Packet switching X.25

Each layer, except the first, has added an additional "envelope" around the message thus:

Layer 4 transport	Ensures that message is in a form that lower layers can handle, e.g. digital binary code. Adds a header indicating nature of sender and recipient. Performs error recovery function
Layer 3 network	Adds user identities and sequence numbering. Adds "signalling" messages (Table 2.4)
Layer 2 *data link*	Adds flow control and check sequence
Layer 1 physical	

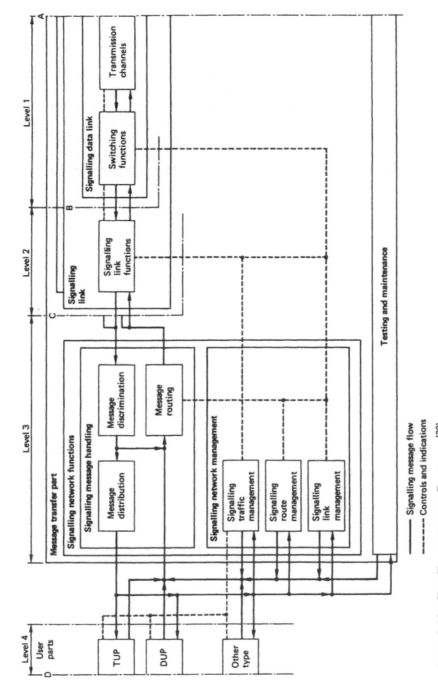

Figure 2.19 Signalling system no. 7 structure[20].

subject of protocol is next addressed directly. Having introduced the concepts, a detailed understanding of examples of the art was gradually built up, starting with coding and considering physical interfaces, link protocols and network protocols. It is no accident that this last brought us back to packet switching, and likewise it is no accident that we found that we have considered the first three layers of the OSI seven-layer model.

Historical assessment of the introduction of the ISDN will no doubt recognize the profound impact that the immediately preceding development of packet-switching systems and their international definition has had upon the definition of the ISDN protocols, particularly in the D channel.

Throughout the chapter, but particularly in the last section on codes and protocols, the information contained in the figures is far more comprehensive than that contained in the text. This is because the diagrams are intended not only to illustrate the text but to serve as definitive introductions to the protocols, such as HDLC and X.25, which they describe.

CHAPTER 3

ISDN tools: intelligent terminals

3.1 Introduction

It has already been indicated that the digital communications network is being introduced for reasons other than the desire to implement the ISDN. PCM time-division multiplexing (TDM) was first introduced to reduce costs and provide improved transmission properties, but it was very soon realized that the combination of PCM TDM transmission and switching with SPC provided a more economic solution for the whole network. Hence, the integrated digital network (IDN) is being put into place for purely economic reasons. A subsidiary effect of combined PCM TDM switching and transmission and SPC is that intelligence is centralized but complexity (of which there is plenty, at least in the analogue-to-digital conversion) is moved to the network periphery (Fig. 3.1). The subscriber's line interface, already the most common, becomes one of the most complex portions of the telephone switching system. With complexity moving towards the subscriber, so too the cost of the system becomes more closely related to the number of subscribers, and the cost of the portion of the system devoted to each subscriber is substantially increased.

The investment in telephone exchange switching equipment must necessarily precede sale of telephone service to the subscriber. Conversely, the subscriber pays for, or rents, the telephone instrument immediately upon receipt. Can the complexity of the line interface therefore be shifted out of the exchange and into the telephone set, thus removing an investment that precedes payment? Of course it can, and the result is a significant economic incentive to create a network that is digital all the way to the subscriber's telephone.

This philosophy has led to the desire to provide digital access at the subscriber's premises. Digital access will increase the cost of the telephone, and thus it is necessary to offset this by making digital access attractive to the subscriber by offering integrated services digital access.

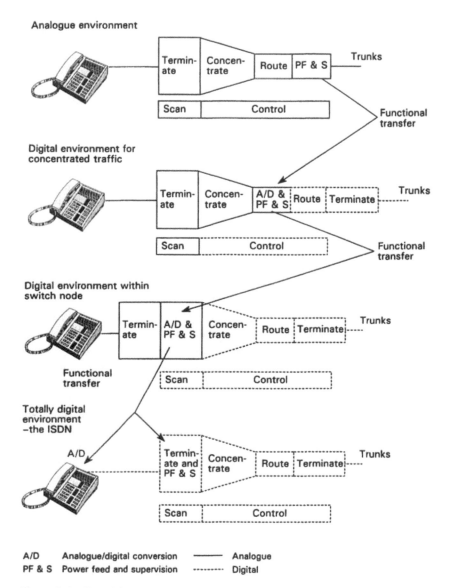

Figure 3.1 Transition stages towards ISDN.

3.2 Integrated services digital access

Chapter 1 indicated the variety of communications services to which the present-day user is already a subscriber: telephone through connection to the PSTN, telex through the telex network, data through private and public circuit-switched data networks and data via the packet-switched data network. With the PSTN becoming digital, all these networks are carrying information in digital form and it remains to devise a universal access method and to standardize the digital formats to allow all services to be available over the same connection, the local telephone distribution network, to the subscriber. The concept of the ISDN is fundamentally a concept of integrated access.

For integrated access to become a reality it was necessary for the telecommunications authorities of the world to agree on the nature of this access by defining the customer interface and the customer-to-network protocols. The resulting requirements and standards and the method of arriving at their definition will be covered later. It is only necessary here to provide details of the accesses defined without discussion at this stage. These accesses are illustrated in Figure 3.2.

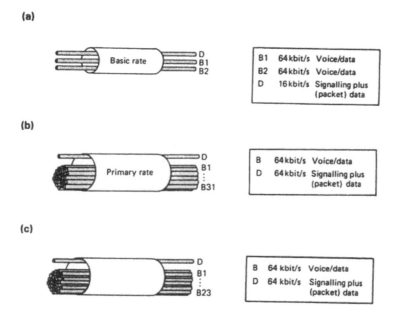

Figure 3.2 ISDN access[21]. (a) Basic ISDN access (I.430). (b) Primary ISDN access 2048 kbit/s (I.431). (c) Primary ISDN access 1544 kbit/s (I.431).

The digital PSTN provides duplex, i.e. four-wire, connections over 64 kbit/s digital channels. Such channels are multiplexed into primary systems, 30-channel systems in most of the world or 24-channel systems in North America and South-East Asia[2]. Access to this network from the subscriber has been defined in two forms:

(a) *Basic access* (Fig. 3.2 a) is equivalent to, but rather more than, access to the 64 kbit/s PCM channel.

(b) *Primary access* (Fig. 3.2 b and c) is equivalent to providing the subscriber with a complete 30-channel or 24-channel system.

The access definitions[22, 23] define two kinds of channel: B channels (64 kbit/s and available for voice or data) and D channels (16 or 64 kbit/s and available for data and signalling only).

The arrangement of ISDN basic access is shown in more detail in Figure 3.3. Perhaps the most attractive feature of it is that the 144 kbit/s capacity of the line is made available to up to eight terminal devices. These may be telephones, facsimile machines or any form of data device, such as a personal computer. Thus, the provision of basic access could remove the need to have a small PABX or key system *and* the need to have a small LAN in many high street shops and offices.

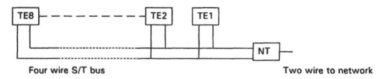

Four wire S/T bus   Two wire to network

Figure 3.3 ISDN basic access.

Having briefly defined access in terms of the communications paths from the subscriber to the network, it is necessary to describe the devices using this access and to define the boundary interface between the subscriber and the ISDN. The accesses shown in Figure 3.2 are certainly a part (perhaps the most important part) of the ISDN, but where does the ISDN end and what is it that requires access?

3.3 The intelligent terminal

Chapter 2 touched on some of the devices and systems that could benefit from the ISDN. Existing personal computers, remote computer terminals, data capture devices and facsimile machines all are currently connected over analogue private and public telephone networks using a variety of dial-up and dedicated private circuits and associated data modems. A modem is a device that converts binary data into pulses of alternating signal frequency within the speech band of 300–3400 Hz. Using a modem data can be passed over the analogue network at speeds of up to 9.6 kbit/s. The speed achievable is very much dependent on the quality of the circuit, and a dial-up connection over the PSTN may only work adequately at speeds of 2.4 kbit/s, although speeds up to 14.4 kbit/s are achievable with modern equipment in some circumstances. In providing such connection the PSTN takes no interest in the compatibility of the devices being connected, so responsibility for determining whether meaningful conversation can take place is left to the user or to the communicating device. We have all probably experienced the annoyance of calling a facsimile machine in error and being able to make nothing of the resulting high-pitched tone we heard.

The existence of such sophisticated terminals and their need to communicate, only inadequately satisfied by the present network, is one of the factors driving the move to the ISDN. These terminals will exist and will be provided by agencies other than the telephone network operators and thus there is a need to define the interface between the terminal and the ISDN and to define it in a fashion that enables pre-ISDN terminals to use it.

3.3.1 The intelligent terminal before ISDN

To understand the terminal and terminal access requirements of the ISDN it will be helpful to consider the terminal as it existed before the ISDN. There are very many such terminals in existence: "dumb" terminals devoted to a particular computer system, telex terminals, terminals to word processor systems and networks, facsimile machines, communicating computers, videotex terminals and personal computers accessing each other or accessing remote databases. Rather than considering specific examples the general specification of such a terminal will be discussed (Fig. 3.4).

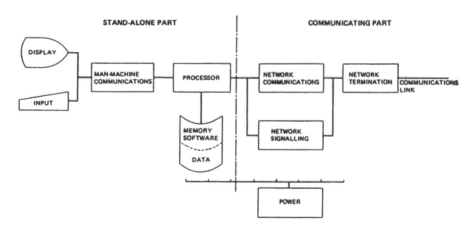

Figure 3.4 General form of terminal device before ISDN.

To the left of the dotted line in Figure 3.4 is a stand-alone device devoted entirely to communicating with the user and executing the user's instructions in a stand-alone mode. If a communications port is provided it will probably be a V.24 interface, but it is not used in the stand-alone application. To the right of the dotted line are shown the functions necessary to communicate. The network link must be terminated and functions provided to convert the data on a V.24 interface into a form suitable for transmission (2.4 kbit/s frequency shift keying for example). In addition, it is necessary to provide signalling protocols at least to alert the far end to a call and to alert the near end to an incoming call. Further signalling is necessary if the communications link is completed via a switched network to provide the necessary instructions to and from the network. Existing terminal devices are powered from the local mains supply, which means that incoming "calls" may be lost when the local power is not switched on.

Contrast this with the existing arrangements in the normal terminal on the PSTN, the analogue telephone (Fig. 3.5). The telephone derives its power from the communications link, so it is always switched on and able to receive calls. The communication protocol is represented by the four-wire to two-wire conversion in the balance network. Signal processing and input and extraction of signals are provided. The combination of balance network and network termination provides an interface to the PSTN that is defined and adhered to worldwide so that all such terminals are compatible.[1]

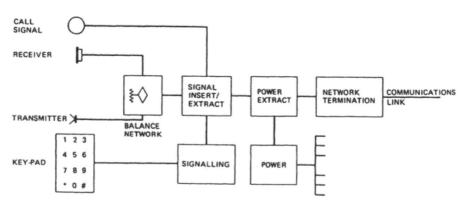

Figure 3.5 Typical telephone before ISDN.

The ISDN terminals must, as far as possible, match the characteristics of the telephone. They must be universal and always alert for incoming calls. They must also have a new feature not required of the telephone, that of indicating their suitability for a particular kind of call (for example a facsimile machine cannot talk to a PC unless the latter is equipped with a fax modem). Telephones are designed and approved to present acceptable interface conditions to the network (they must not impede the way in which the network operates). With more complex features the terminal also must not adversely affect the network, and must interwork with other similarly complex terminals.

---

1. But not portable, an American phone may not work if plugged into a UK telephone line and in any case the plug is different.

3.4 The ISDN terminal: concept

At a reasonably early stage in the definition of the ISDN requirements the terminal interfaces were defined at the reference points shown in Figure 3.6. The definitions make use of the concepts of open systems interconnection (OSI) which is treated in the next chapter. At this introductory stage the interfaces will be discussed in more general terms. Figure 3.6 shows the terminal divided into functional groupings. It is only by separating functions conceptually in this way that it is possible to identify and specify the interfaces between them. Before describing the interfaces it is appropriate to describe the functional groups. This initial treatment will be confined to the functions of basic access.

    (a) *Network termination 1 (NT1).* This is the group of functions that terminate the transmission line. As such, it is seen in the CCITT recommendations as belonging to the network provider, the owner of the transmission line. The functions of the NT1 are described as: line transmission termination; line maintenance and performance monitoring; timing; power transfer, i.e. extracting power from the line to drive at least the "wake-up" portion of the terminal; parts of the multiplexing functions;

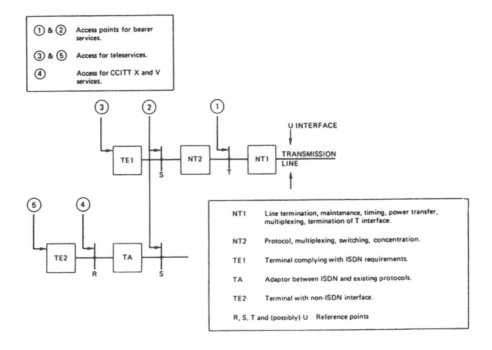

Figure 3.6 Customer access to services supported by an ISDN[24].

(b) *Network termination 2 (NT2)*. This is the group of functions that give the terminal its particular "character". An NT2 could be a PABX if access is primary, a LAN or a terminal controller. The functions of the NT2 are described as: protocol handling or handling that part of the protocol associated with information transfer across a network; the higher level parts of the multiplexing function; switching and concentration functions; maintenance functions; termination of the S interface, which may include multidrop termination and associated contention resolution functions; interface functions to the S and T interfaces.

The NT2 functional group may be more or less complex depending on the application. The range extends from the quite complex function of a PABX down to relatively simple functions required for a time-division multiplexer. In specific, simple cases all the functions may be adequately performed by NT1, and NT2 becomes merely physical connections.

(c) *Terminal equipment (TE)*. These functional groups are broadly similar to the equipment shown in the left-hand part of Figure 3.4. This is the device itself. It could be a digital telephone, a CAD/CAM workstation or a computer terminal.

- *Terminal equipment type 1 (TE1)* complies with ISDN user–network interface recommendations and therefore supports interface S.
- *Terminal equipment type 2 (TE2)* supports the same functions but does not comply with the ISDN user–network interface recommendations. It must therefore interface with the ISDN access via a *terminal adaptor.*
- *Terminal adaptor (TA)* converts the non–ISDN interface functions into ISDN-acceptable form at reference points S or T.

The above descriptions have largely revealed the nature of the interfaces and, indeed, the CCITT recommendations offer no further explanation of the reference points than a similar description of the functional groupings that they separate. To summarize, however, we will list the nature of the reference points shown in Figure 3.6.

Reference point T. (The T is for terminal.) This separates the network provider's equipment from the user equipment. It provides a standardized interface between equipment sending and receiving validating and timing information to the network and to terminal equipment devoted to the use of this information. Reference point T interfaces with the passive bus in a multiterminal basic access such as that shown in Figure 3.3.

Reference point S. (The S is for system.) This separates the user terminal equipment from the network functions of the terminal.

Reference point R. (The R is for rate.) This provides a non–ISDN interface between non-ISDN-compatible user equipment and adapter equipment. Such an interface may well comply with one of the CCITT X series interface recommendations.

Figure 3.6 indicates another interface at the transmission line:

Reference point U. (The U is for user). Were this not indicated the reader might well have been surprised by the absence of reference to any such interface. The only explanation for its absence in the CCITT recommendations is note 2 to paragraph 3.2 of Recommendation I.411[25], which reads:

> Note 2. There is no reference point assigned to the transmission line, since an ISDN user–network interface is not envisaged at this location.

This less than comprehensive statement conceals a continuing subject of controversy that deserves brief treatment here.

The First Amendment to the Constitution of the United States severely restricts the government's role in areas related to information gathering, processing and dissemination. The provision of telecommunications services, however, has always been subject to considerable government regulation, largely by the Federal Communications Commission (FCC). The gradual merging of these functions (information exchange and communications) over the past 20 years has led the FCC to consider and define the relationships between information and telecommunications, as well as what should be regulated and what ought not to be regulated. Important definitions were reached as a result of the Computer enquiries, most importantly Computer II (1981). This decision defined two kinds of telecommunications service: basic service, which must be made available to everyone, and enhanced service. This issue will be discussed later (see §8.2.4). Computer II also defined a distinction between network equipment and customer premises equipment (CPE). The latter must be available to users freely and independently of the network provider.

The emergence of the concepts of the ISDN led the FCC to consider their impact on the Computer II decisions. This was done by means of an enquiry on the ISDN, the conclusions of which were published in ISDN First Report in 1984. In this report it is acknowledged that digital network carrier terminating equipment (NCTE) has already been identified as CPE in the terms understood by Computer II and therefore that the concept of the CCITT-defined NT1 being part of the network is contrary to Computer II. Thus, "our NCTE decision requires that there be established an interface to the loop to which CPE (including any NT1 device ...) may be connected". The U interface lives, at least in the USA.

3.5 The ISDN terminal: design

Having defined the terminal in terms of functions and interfaces it is now necessary to consider how such a terminal might be described to an engineer entrusted with its design. In the previous section no distinction was made between basic rate and primary rate terminals; for the purposes of this description a basic rate access will be assumed. It will also be assumed that it is a terminal made up of TE1, NT2 and NT2 (the top half of Figure 3.6).

The design brief is summarized in Figure 3.7 and the other figures to which Figure 3.7 refers. An immediate problem is the absence of a CCITT recommendation for the U interface. Physically this is a pair of wires. The information protocols used are considered to be the affair of the administration, the network provider. In fact, however, the nature of the interface is dictated by the recommendations for interfaces S and T and the requirements of the V interface between the NT and the exchange itself. Thus, the protocol for line signalling is dictated by that for the S/T interface, as shown in Figures 3.9 and 3.10.

At the S (or T) interface, a four-wire physical configuration is specified (Fig. 3.8) and a pseudoternary communications protocol is used, illustrated in Figures 3.9 and 3.10. These figures illustrate the limiting interfaces within which the designer must work. Allied with these, the design specification must include features such as the following:

(a) At least the "wake-up" portion of the terminal must derive power from the exchange over the two-wire line. The exchange includes a feature to send a reversal should a fault cause limitation of power. The terminal must reduce its consumption to an absolute minimum on receipt of this reversal signal.
(b) Facilities for "wake-up" and subsequent synchronization in both directions of transmission.
(c) Facilities for maintenance signals. These will, for example, complete testing loops towards both interfaces in the NT1.
(d) Space for information for two B channels and one D channel plus word synchronization.
(e) Frame synchronization.
(f) Arrangements to allow duplex transmission over the two-wire line.

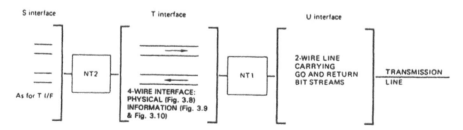

Figure 3.7 Basic rate ISDN terminal design.

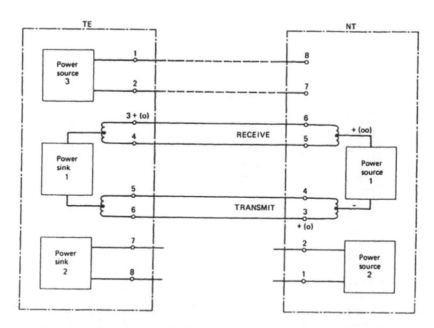

Figure 3.8 Reference configuration for signal transmission and power feeding in normal operating mode[22]. Note that the numbering of leads does not imply any assumption on pin allocation or physical connectors. Maintenance of polarity on a wired pair cannot be guaranteed in all cases. This must be taken into consideration for terminals drawing power from access leads 7 and 8. o, Polarity of positive pulses; oo, polarity of power.

(g) Contention for the line may occur where several TE1s are connected to the S interface. Contention is avoided by use of a technique known as carrier sense multiple access with contention resolution (CSMA-CR). All bits received by the NT over interface S or T in the D channel are echoed in the E channel, the D-echo channel. When the D channel is quiescent, the NT sends all ones in the E channel. A terminal detecting eight or nine consecutive 1s in the E channel may seize the link as such a sequence is not allowed by the normal D channel protocol (HDLC). The terminal continues to match its own sent information against the returned D echo and knows, if there is a mismatch, that collision has occurred. On collision detection, the terminal must stop and try again.

(h) The S and T interfaces use a pseudoternary code where 1 is zero voltage and, in general, successive "0"s are alternately positive and negative. This allows a degree of checking of the information sent. The algorithm used to determine the polarity of "0"s is as follows:

"The first binary zero after the framing balance bit is at the same polarity as the framing balance bit. Subsequent binary zeros must

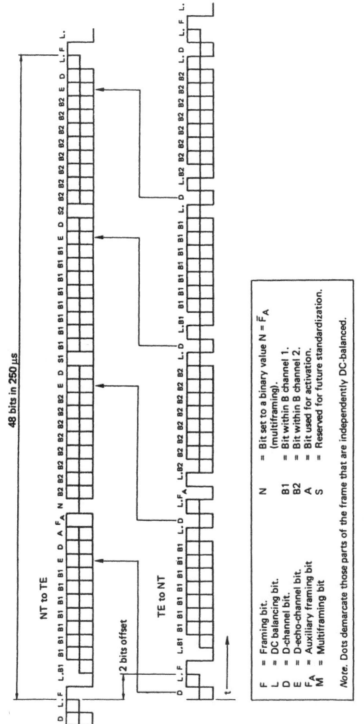

Figure 3.9 Frame structure at reference points S and T[22]. Dots demarcate those parts of the frame that are independently DC balanced.

F = Framing bit.
L = DC balancing bit.
D = D-channel bit.
E = D-echo-channel bit.
F_A = Auxiliary framing bit
M = Multiframing bit

N = Bit set to a binary value N = $\overline{F}_A$ (multiframing).
B1 = Bit within B channel 1.
B2 = Bit within B channel 2.
A = Bit used for activation.
S = Reserved for future standardization.

Note. Dots demarcate those parts of the frame that are independently DC-balanced.

Figure 3.10 Pseudoternary code example of application[22].

alternate in polarity. A balance bit is a binary zero if the number of binary zeros following the previous balance bit is odd. A balance bit is a binary one if the number of binary zeros following the previous balance bit is even".[22]

3.6 The ISDN terminal: reality

With a great deal of imagination, our designer might come up with a design for an NT1 that resembles the block diagram shown in Figure 3.11. The designer of Figure 3.11 has, in fact, been privy to much more information than we provided in the previous section including, for example, a requirement to scramble the information.

There is provision for a wake-up signal, a 7.5 kHz analogue tone in either direction. The recognition circuitry for this signal in the NT1 is not shown. The line power is used to power the wake-up circuitry only, and, again, the power extraction equipment is not shown. The two B- and one D-channels towards the U interface are scrambled, converted to ternary line code (Fig. 3.12), filtered and presented to line.

The surprise appearance of a digital-to-analogue conversion function at this point is because of a misnomer. This conversion is a pulse-shaping function to limit the higher frequency content sent over the line and thus improve the cross-talk and noise immunity characteristics of the local ISDN network.

The diamond symbol and the echo-cancelling network represent the four-wire to two-wire conversion features of the interface. In analogue telephony, echo was only a problem in long-distance circuits in which the transmission delay was sufficient to make the returned echo of the sent signal, reflected because of imperfections in the termination circuitry, obtrusive. This effect was cured by an echo suppression feature. In data transmission, echoes that distort the received information are objectionable if they make it difficult to distinguish between the coding levels. In recent years the availability of digital LSI circuit techniques has made echo cancellation, subtracting the signal sent from the signal received, a practical alternative to the less complex solution of echo suppression. In the digital hybrid this same technique can be used to separate the signal sent from the real signal received and the received echo of the signal sent. The remainder of the receive circuitry, apart from the thresh-

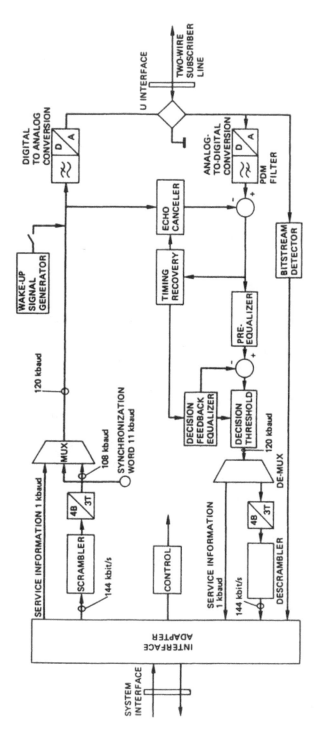

Figure 3.11 Block diagram of the line-termination/network-termination 1 VLSI circuit. Redrawn with permission from ref. 26.

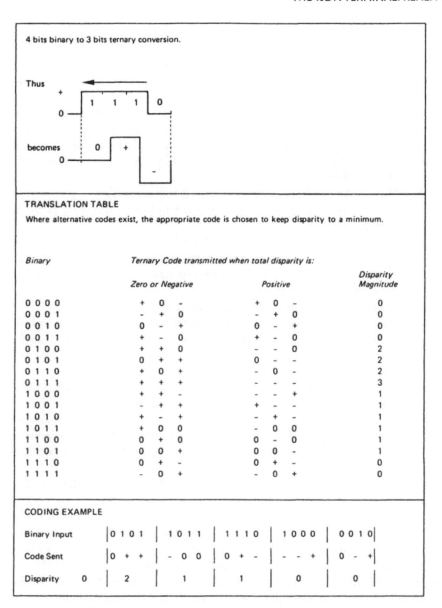

4 bits binary to 3 bits ternary conversion.

TRANSLATION TABLE

Where alternative codes exist, the appropriate code is chosen to keep disparity to a minimum.

Binary	Ternary Code transmitted when total disparity is:						Disparity Magnitude
	Zero or Negative			Positive			
0 0 0 0	+	0	–	+	0	–	0
0 0 0 1	–	+	0	–	+	0	0
0 0 1 0	0	–	+	0	–	+	0
0 0 1 1	+	–	0	+	–	0	0
0 1 0 0	+	+	0	–	–	0	2
0 1 0 1	0	+	+	0	–	–	2
0 1 1 0	+	0	+	–	0	–	2
0 1 1 1	+	+	+	–	–	–	3
1 0 0 0	+	+	–	–	–	+	1
1 0 0 1	–	+	+	+	–	–	1
1 0 1 0	+	–	+	–	+	–	1
1 0 1 1	+	0	0	–	0	0	1
1 1 0 0	0	+	0	0	–	0	1
1 1 0 1	0	0	+	0	0	–	1
1 1 1 0	0	+	–	0	+	–	0
1 1 1 1	–	0	+	–	0	+	0

CODING EXAMPLE

Binary Input	0 1 0 1	1 0 1 1	1 1 1 0	1 0 0 0	0 0 1 0
Code Sent	0 + +	– 0 0	0 + –	– – +	0 – +
Disparity 0	2	1	1	0	0

Figure 3.12 4B3T coding.

old circuitry, is the reverse equivalent of the send circuits.

The perceptive reader will have compared Figure 3.9 with the kbit/s and k-baud figures on Figure 3.11 and realized that not all the channels identified in Figure 3.9 are sent across the U interface in Figure 3.11. In this particular realization only B1, B2, D, F and FA, A and N channels are included.

3.7 Line coding

The problem of sending digital information over a two-wire subscriber's local line is a subject in its own right that deserves a few paragraphs of treatment here.

Line coding is discussed further in Chapter 9 in connection with the clear channel capability problem. There the discussion is about four-wire trunk connection between network nodes. Here we are confining ourselves to the possibly more difficult problem of digital, duplex communication over the two-wire subscriber's line.

In all line-coding techniques there is a need to provide for:

(a) Continuously alternating signal with low bandwidth and good facilities for timing recovery.
(b) Low attenuation and therefore long lines (> 5 km).
(c) Protection from noise interference, in particular near-end cross-talk, the transmitted signal interfering with the received signal at or near the terminal from which it is transmitted.
(d) Low complexity to the echo cancellation function.
(e) Fast convergence of echo cancelling. On switch-on there is a perceptible time delay before the echo-cancelling routine gets into its stride. On subscribers' systems that discontinue transmission during idle periods, this delay must be minimal.
(f) Low intersymbol interference (ISI), the delayed remains of one symbol affecting the waveform of the next and subsequent symbols.
(g) High signal-to-noise ratio (SNR). This is greatly affected by the number of decision levels in the signal. Pure binary signals with just two decision levels will, other things being equal, have a higher SNR than multilevel signals.

It is possible to distinguish between line-coding methods that involve watching and converting the bit stream "as it passes", called linear coding methods, and those that require re-encoding the bit stream by means of a look-up table, called block coding methods. AMI and HDB3 (Fig. 9.4) are examples of linear codes, although 4B3T (Fig. 3.12) is an example of a block code. Another very simple linear code is biphase, which sends a full cycle of a square-wave signal for each bit but distinguishes between "1" and "0" by sending "1" as the opposite phase of the signal used to represent "0".

Whereas early line signalling systems used biphase, the two main contenders for CCITT recognition have been AMI and 4B3T. The decision taken for North American standardization introduces a fresh contender, 2B1Q, illustrated in Figure 3.13.

The linear codes with just two decision levels tend to send information to the line at the same rate as it is received in binary form (160 kbit/s for the CCITT basic access). The multilevel codes, by contrast, tend to reduce the sending rate because more information redundancy is available in the code. In general,

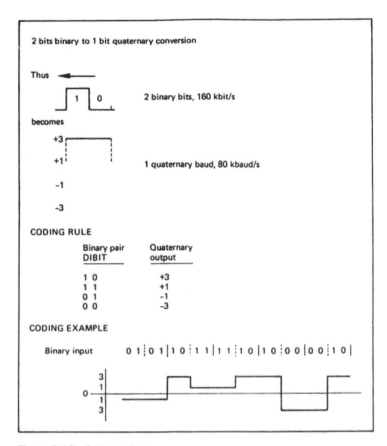

Figure 3.13 2B1Q coding.

therefore, such codes are more attractive although they require greater complexity to combat cross-talk and intersymbol interference. 4B3T, for example, uses a 120 kbaud/s line rate and 2B1Q achieves 80 kbaud/s but at the expense of lower SNR (more decision levels) and more ISI (lower baud rate).

For a more extended discussion of line coding the reader is recommended to read Chapter 4 of reference 27.

3.8 The CCITT access recommendations

So far ISDN access has been discussed while avoiding continuous reference to the CCITT recommendations that are the definitive specification source. Nevertheless, the recommendations have been referred to frequently and some of their present inadequacies have been identified. Having reached an understanding of what is required of ISDN access it is timely to consider how the recommendations express these requirements and the areas that they fail to address.

The primary objective of the CCITT has been to ensure that the ISDN, when it is deployed, will be just as ubiquitous as the existing PSTN that it enhances. One concept used by the CCITT to express this universality has been that of portability: an ISDN terminal should be capable of being plugged into the ISDN anywhere in the (ISDN) world. This is more than the PSTN has been able to achieve. One could carry a telephone instrument around the world and achieve some sort of service, but the plug is not a universal standard and the line-terminating impedances are not world standards so that connection would be a problem and transmission quality erratic. Nevertheless, this is the objective chosen for ISDN access. Laudable, difficult and perhaps unlikely to be completely achieved.

It is not an aim of this book to repeat more than a necessary minimum of the CCITT recommendations, nor is it an object to provide a substitute for the recommendations. Any reader working in ISDN topic areas or affected by the ISDN cannot, and will not wish to, avoid becoming familiar with the recommendations.

The bulk of the information on the ISDN is contained in the Blue Books (1988 Recommendations) in three volumes, Volume III, Fascicles III. 7, III.8 and III.9. However, these volumes refer to other volumes for some of the most essential recommendations. This is illustrated for the layer 2 and layer 3 user–network interfaces in Table 3.1. The table also indicates that the versions issued in the 1988 Blue Book are now out of date, revision 1 of all these documents having been issued in 1993.

For our present purpose we are interested in the I.400 series recommendations and comments on these will form the remainder of this chapter.

Table 3.1 ISDN recommendations from CCITT.

Title	Recommendation	Refers to Recommendation	Date	Fascicle
ISDN user–network interface data link layer – general aspects	I.440[28]	Q.920[33]	1993	VI.10
ISDN user–network interface, data link layer specification	I.441[29]	Q.921[34]	1993	VI.10
ISDN user–network interface layer 3 – general aspects	I.450[30]	Q.930[35]	1993	VI.11
ISDN user–network interface layer 3 specification for basic call control	I.451[31]	Q.931[36]	1993	VI.11
Generic procedures for the control of ISDN supplementary services	I.452[32]	Q.932[37]	1993	VI.11

3.8.1 I.410: general aspects and principles

This recommendation indicates the kind of interfaces envisaged (Fig. 3.14), the characteristics that must be specified and the interface capabilities to be supported. The capabilities are stated to include multidrop (several terminals contending for the same access), choice of features (bit rate, coding method, etc.) on a call-by-call basis and facilities for compatibility checking (fax cannot talk to viewdata).

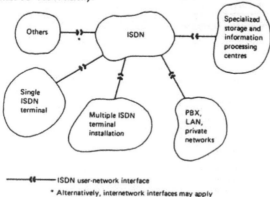

———◀◀——— ISDN user-network interface

* Alternatively, internetwork interfaces may apply

Figure 3.14 ISDN user–network interface examples[38].

3.8.2 I.411: reference configurations

This recommendation provides the definitions of NT1, NT2, etc. and reproduces and expands upon Figure 3.6 to illustrate the kinds of physical configurations envisaged and the implementations of NT functions expected. Figure 3.15 is reproduced here as an example.

An interesting additional possibility introduced by I.411 is that of hybrid ISDN and analogue access to the same transmission line. No attempt will be made here to discuss how this might be achieved, but it is a necessary feature in some instances that can be foreseen. Consider, for example, the power fail transfer feature provided (mandatorily in the UK) in PABX equipment. This feature switches exchange lines directly to nominated extension telephones should the PABX power supply fail. For a PABX with ISDN access it would be important to ensure that these extensions did not depend on local power, and this might only be possible by making the extensions analogue.

3.8.3 I.412: interface structures and access capabilities

In this recommendation[21] the channel types envisaged for ISDN are defined and the interface structures using these channels are described. A summary of the content of the recommendation is provided here, reinforcing references

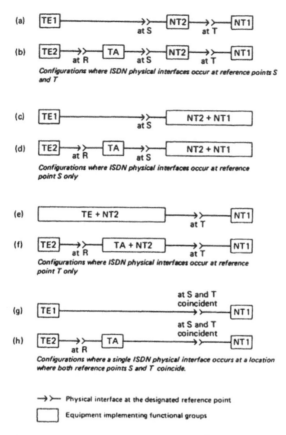

Figure 3.15 Examples of physical configurations[25].

made earlier in the chapter and providing a simplified reference glossary. These interface structures are largely theoretical and not exhaustive. There are many eventualities for which they do not cater. For example, a user requiring several low bit rate channels has only the option of using several basic accesses or primary access, in either case wasting most of the available bandwidth.

3.8.3.1 Channel types
The following channel types are available.

(a) B channel (64 kbit/s accompanied by timing). The B channel is intended for voice or data, possibly submultiplexed (portions of) wideband voice. Circuit switching, packet switching and semipermanent connections are supported. Voice encoding must be to Recommendation G.711[40]; data obeys Recommendation X.1 and/or X.25. For circuit switching only the complete 64 kbit/s B channel is switched.

(b) D channel (16 kbit/s basic access or 64 kbit/s primary access). The D channel is primarily intended for signalling and uses a layered protocol.

72

It may also be used for teleaction (telemetry) and packet-switched data.
(c) H channels: H0 (384 kbit/s); H1 and H11 (1.536 Mbit/s, for 1.544 Mbit/s primary rate); H12 (1.920 Mbit/s, for 2.048 Mbit/s primary rate).

Higher bit rate H channels may be specified later. H channels are intended for a variety of user information streams (fax, video, etc.) but not signalling.

3.8.3.2 Interface structures

The fundamental structures were illustrated in Figure 3.2. These are repeated together with the other structures identified in Recommendation I.412 in Figure 3.16.

I.420 and I.421[41, 42] describe the basic and primary interfaces but provide no additional information.

Recommendation I.430 describes basic interface layer 1[22]. This and the remaining recommendations descend to a level of detail where it is necessary to use the concepts of the OSI model. The concept of layers was introduced in

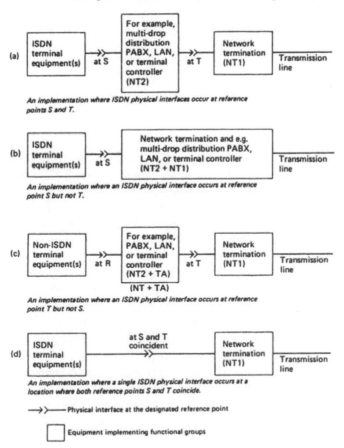

Figure 3.16 Interface structures.

73

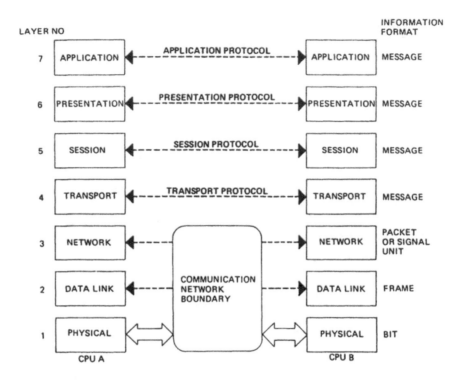

Figure 3.17 OSI model.

Chapter 1 but a full discussion of the OSI model is reserved for Chapter 4. Here it is enough to recall the layer functions illustrated in Figure 1.7, which is reproduced here as Figure 3.17 for convenience.

Recommendation I.430[22] defines the layer 1 characteristics of the user–network interface, reference points S or T (Fig. 3.6). Layer 1 must support transmission over a balanced metallic pair for each direction of transmission capable of 192 kbit/s. The layer 1 services offered to layer 2 are: transmission capability for B and D channels together with timing and synchronization; signalling to activate and deactivate terminal equipment; signalling to allow multiple terminals to contend for the common D channel; and signalling for maintenance functions. Recommendation I.430[22] describes the various permitted wiring configurations for point to point (one terminal per access) and point to multipoint (up to eight terminals per access).

The description of the functional characteristics makes use of Figures 3.9 and 3.10 and the remainder of the recommendation describes the interface procedures, maintenance, electrical characteristics and power feeding. Using I.430 and its annexes, it is possible to design and build a layer 1 ISDN user–network interface.

Recommendation I.431, primary rate user–network interface – layer 1 specification[23], provides a description of the primary rate interface in a little less detail than the basic interface was treated by I.430. Only point-to-point working is envisaged, but two primary rates are specified: 1.544 Mbit/s for North America and South-East Asia and 2.048 Mbit/s for world use. The functional characteristics call for support of B, H0, H1 or D channels (the D channel being 64 kbit/s).

For the 1.544 Mbit/s rate a 24-frame multiframe is defined as mandatory (Fig. 3.18 and Table 3.2) The 2.048 Mbit/s rate interface is identical to the

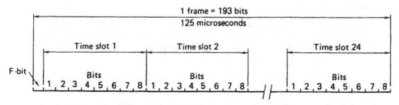

Figure 3.18 1.544 Mbit/s primary access structure. Frame structure of 1.544 Mbit/s interface.

Table 3.2 Multiframe structure of 1.544 Mbit/s interface (Table 5 of Recommendation I.431[23]).

Multiframe frame number	Multiframe bit number	F bits		
		Assignments		
		FAS	Maintenance	Error check
1	1	–	m	n[*]
2	194	–	–	e_1
3	387	–	m	–
4	580	0	–	–
5	773	–	m	–
6	966	–	–	e
7	1159	–	m	–
8	1352	0	–	–
9	1545	–	m	–
10	1738	–	–	e_3
11	1931	–	m	–
12	2124	1	–	–
13	2317	–	m	–
14	2510	–	–	e_4
15	2703	–	m	–
16	2896	0	–	–
17	3089	–	m	–
18	3282	–	–	e_5
19	3475	–	m	–
20	3668	1	–	–
21	3861	–	m	–
22	4054	–	–	e_6
23	4247	–	m	–
24	4440	1	–	–

[*]The appearance of the symbol "n" as well as the "m" bit in multiframe bit 1 is not of course another value; there can only be one value for a single bit. "n" probably refers to §2.1.3.1.2 of Recommendation G.704 and indicates the start of CMB N of the cyclic redundancy check (CRC-6 Message Block N).

Table 3.3 2.048 Mbit/s primary access arrangements in channel 0 (Table 4a of Recommendation G.704[43]).

				Bit number				
Alternate frames	1	2	3	4	5	6	7	8
Frame containing the frame alignment signal	S_i*	0	0	1	1	0	1	1
				Frame alignment signal				
Frame not containing the frame alignment signal	S_i*	1†	A‡	S_{a4}	S_{a5}	S_{a6}	S_{a7}	S_{a8}
				See note 1 below				

*Reserved for international use. One particular use is for the CRC-4 procedure.
†Fixed at "1" to avoid simulations of the frame alignment signal.
‡A = remote alarm indication; "1" indicates an alarm.
1. Bits used for various applications; transcoders to G.761, S_{a4} may be used as a data link for operations, maintenance and performance purposes; failing these uses they are available for national use. Set to "1" when not otherwise used.

CEPT 30-channel system[2] except that the function of the bits in channel 0 of the frame are defined in more detail (Table 3.3).

The remaining I-series recommendations deal with the higher level layers of the interfaces and are beyond the scope of the treatment suitable in this chapter. For completeness, however, the titles are listed here:

| Layer 2: | I.440[28] | data link layer – general |
| | I.441[29] | data link layer – specification. |

| Layer 3: | I.450[30] | layer 3 – general |
| | I.451[31] | layer 3 – specification. |

Miscellaneous:	I.460[44]	multiplexing rate adaptation and support of existing services
	I.461[45]	support of X.21 and X.21bis. DTEs
	I.462[46]	support of packet-mode terminals
	I.463[47]	support of V series DTEs
	I.464[48]	multiplexing, rate adaptation and support of existing services for restricted 64 kbit/s transfer capability.

Chapter summary

The discussion commenced with the economic arguments that encourage the digital network and the variations on these arguments that make it sensible for both the network provider and the subscriber to desire to furnish this access at the subscriber's premises. Integrated digital access is a more complex function than PSTN access, and the forms of access agreed internationally were introduced without discussion.

Integrated digital access is predicated upon the existence of an intelligent terminal capable of making use of such sophisticated communications media. The discussion of terminals started from the existing devices, including the telephone, and moved on to elaborate the conceptual requirement of an ISDN terminal. This conceptual discussion introduced the concepts of interfaces and functional groupings used in the CCITT I-series recommendations on the ISDN. A short excursion into American communications legislation explained the difficulty of the U interface between the ISDN terminal and the two-wire subscriber's line.

These concepts were then put into practice by introducing the problems of ISDN terminal design. In making sense of this concrete realization considerably more material from the I-series recommendations was introduced.

Having again delved into the I-series recommendations the chapter ended with a more structured summary of their contents.

CHAPTER 4

ISDN tools: common channel signalling and interoperation; open systems

4.1 Common channel signalling (CCS)

Digital transmission and switching allied to SPC provided a very powerful tool with which to realize integrated networks. It is a tool, however, that is largely ineffective unless the switching machines, the exchanges, can communicate freely. Traditionally interexchange communication was limited to the information required to set up, charge, supervise and release calls, and this signalling was conducted over the same path as the conversation to which it related. This is channel-associated signalling. Included in channel-associated signalling is the slightly different arrangement in digital systems, in which the signalling bits are channel associated but passed via a different bearer channel (channel 16 in 30-channel PCM).

The tremendous increase in long-distance and international telephony caused by the advent of direct dialling and undersea cables using repeaters and, later, satellite communications, led to the concept of providing signalling more effectively for large numbers of conversational circuits over a common channel. The appearance, at roughly the same time, of SPC made such CCS a practical proposition. At about the same time as CCS and SPC were being mooted, far-seeing engineers, notably in France, were suggesting that exchanges be controlled over an area by just a few SPC processors[49]. Such a concept also demanded CCS.

CCITT CCS system no. 6 (#6) was defined as a result of these suggestions and the recommendations were first published in reference 6. #6 was an analogue system[1] and its widespread introduction was overtaken by events in the form of the rapid introduction of digital transmission. There seemed to be little point in converting digital control signals into analogue form within the speech band only to code the complete speech band into a 64 kbit/s digital form. #6 signalling was used extensively to upgrade the features of the US trunk network and it persisted in the US into recent times in a digital form created by "stuffing" the 28-bit #6 "word" with four extra bits to make it

---

1. Analogue in the sense that it was intended for use over analogue links using modems. The messages are, of course, in digital form.

78

compatible with the octal word of the digital transmission system. It is not intended, in this book, to provide complete coverage of CCS. The treatment is limited to that necessary to understand the ISDN implications of CCS. Reference 6 treats #6 standards, whereas reference 2 provides a general introduction to #6 and a more detailed treatment of #7.

The widespread introduction of digital transmission and, shortly thereafter, the introduction of digital SPC switching made it imperative to develop systems of digital CCS. The immediate result was CCITT signalling system no. 7 (#7), conceived as a digital replacement of #6.

The first part of this chapter will be devoted to digital CCS. It might have been supposed that this would involve the consideration of one system, #7, but for various reasons, not all of them necessarily sensible, this is not the case. The study will begin with a consideration of #7 and, recognizing that this has been treated in some detail elsewhere (reference 2 is only one of many such treatments), the discussion here will be limited to that necessary and essential to the ISDN.

4.1.1 CCITT signalling system no. 7 (#7)

Signalling system no. 7, conceived as a digital CCS system and devised to eliminate many of the problems associated with #6, is a variable word length system that uses levels of abstraction to divorce the requirements of the transmission medium from those of the communicating processes and processors and to provide a signalling system independent of the nature of the messages being transported. Figures 4.1, 4.2, 4.3 and 4.4 illustrate the principles of the system defined in CCITT Recommendations Q. 701–714[2]. At level 4, the most remote level from the signalling link, telephone user part, data user part, ISDN user part and operation and maintenance user part are defined as alternatives.

A glance at the format of #7 signal units in Figure 4.4 is enough to note the similarity between #7 protocol and the protocol discussions of Chapter 2. It is debatable whether CCS gave the idea of "enveloping" to the data industry and hence to OSI or vice versa. Certainly, the same principle is in use. Indeed,

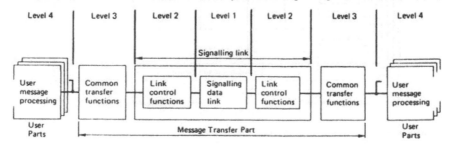

Figure 4.1   System no. 7, functional division concept.

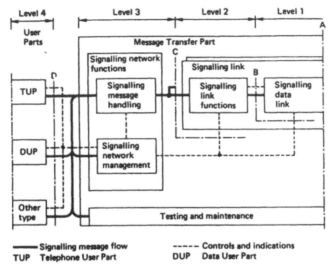

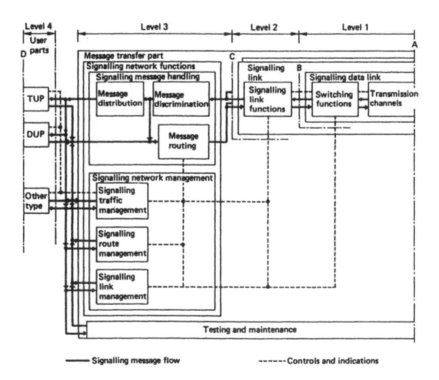

Figure 4.2 System no. 7, conceptual architecture.

Figure 4.3 System no. 7, structure in more detail.

80

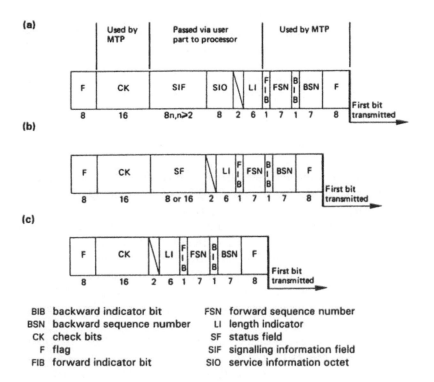

Figure 4.4 System no. 7, basic signal unit formats. (a) Basic format of a message signal unit (MSU). (b) Format of a link status signal unit (LSSU). (c) Format of a fill-in signal unit (FISU).

the telephone industry thought that in defining #7 this was the only signalling system that would be necessary in digital SPC[2] and that the #7 signalling facility could be made freely available for the transparent carriage of data. It was such ideas mooted at the inception of the #7 discussions that gave support to the idea of an ISDN.

The information-carrying capacity of a #7 signalling link is, to a large extent, dependent upon the nature of the error-correcting protocol chosen. Assuming that the "basic" method is used, the capacity of a 64 kbit/s channel devoted to #7 will be significantly in excess of the signalling traffic associated with the control of, say, 1000 traffic circuits. When one considers the necessary provision of alternative CCS routes for security reasons then it is evident that, without special provision, the telephone network can offer considerable spare CCS capacity for other uses.

This spare available capacity will first be used for administration data requirements from network operations through management to non-

---

2. Apart, that is, from the channel-associated signalling mechanism defined for channel 16 (30-channel) and bit 8 of channels 6 and 12 (24-channel D2) (Chapter 3 of reference 2).

telephonic data. It is equally available as a transparent error-controlled data transport mechanism providing public data network capability virtually "free" as part of the digital telephony network. There is, of course, no limit to the number of channels dedicated to CCS purposes and therefore no real capacity limitation to data transport over the telephony network.

4.1.2 #7 reservations

It was the intention of the designers that #7 signalling provide a completely transparent signalling mechanism. Anything included in the signalling information field (SIF) of a message signal unit (MSU) can be transported end to end from a source to a destination indicated by the address label that forms the first part of the SIF. Apart from consideration of this address and processing the complete message to determine the check bits, the network need know nothing of the contents of the message and the communicating entities need know nothing of the network. It is arguable, therefore, that, given #7 signalling, no other mechanism is necessary to provide any form of data communication across the network.

Network providers have seen the matter in a different light. A primary function of #7 is to convey network control, network management and network administration information, and this includes charging information. In view of this, many administrations are reluctant to make #7 facilities available to their users for fear that the user may find ways of interfering with the network. Whether these fears are well founded is matter for debate, but the fear is there and has generated a variety of digital CCS systems whose relationship to #7 must now be considered. A more justifiable reason for limiting the use of #7 used to be that it is a powerful system for use between powerful communicating processors and much of this power is wasted in providing communication between relatively simple user devices. As user devices have become more complex, this argument is no longer true. The degree of sophistication of the ISDN user CCS DSS 1 is at least equal to that of #7.

The reluctance to use CCS for anything other than the management of the communications network has been reinforced by the succession of network disasters, mainly in the US network, in recent years. Not all of these were because of the CCS network itself, although one of the major disasters occurred when the CCS software was being upgraded in part of the network. But all of them were magnified by the widespread use of CCS, so that a fault in a small portion of the network affected large areas up to and including the whole continent. None of these incidents is attributable to misuse of the signalling system itself – they have all been because of poor network and exchange management – but they have enhanced the perception of the importance of the CCS network and therefore increased the reluctance to use it for "normal" (data) telecommunications conversational purposes.

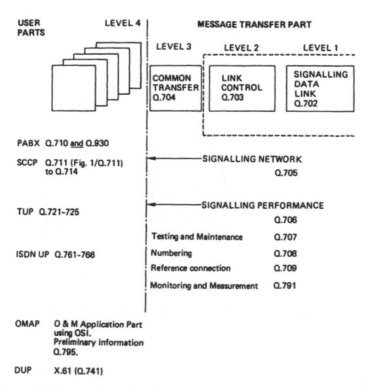

Figure 4.5 System no. 7, signalling parts.

These reservations of the network providers about making #7 capacity available to the user have influenced the negotiations leading to definition of the #7 user parts. A consideration of the variety of digital CCS must therefore start with an account of the #7 user parts presently defined. Figure 4.5 indicates the existing parts of #7 and their definition recommendations.

4.1.3#7 Message transfer part (MTP)

So far in this chapter the text and references cited, in particular reference 2, have described the #7 MTP as a transparent medium for transporting good, error-free data. Figure 4.6 encapsulates the principles of one of the several error correction methods made possible by the forward and backward sequence number and bit mechanism that is perhaps the most important feature of the MTP.

Figure 4.7, however, demonstrates how the MTP must look further within the #7 "envelope" to establish the addresses of the conversing parties and so enable performance of its switching function. This would seem to be no real problem, and certainly preferable to extending the message by a repetition of

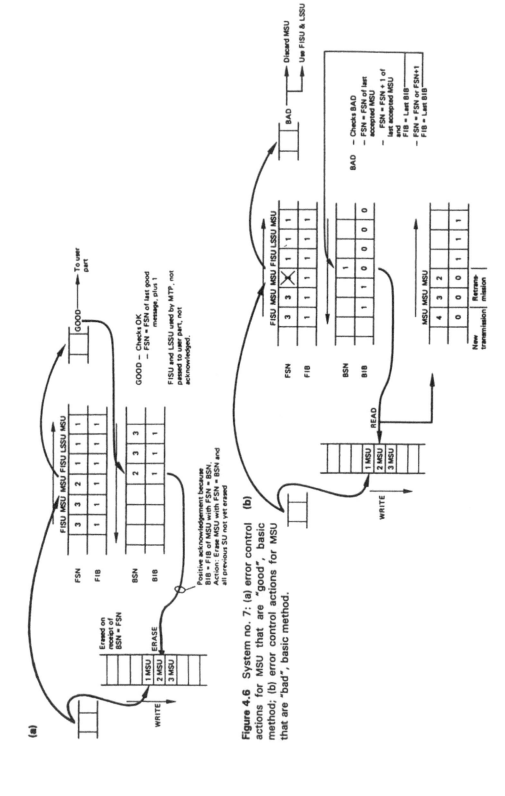

Figure 4.6 System no. 7: (a) error control actions for MSU that are "good", basic method; (b) error control actions for MSU that are "bad", basic method.

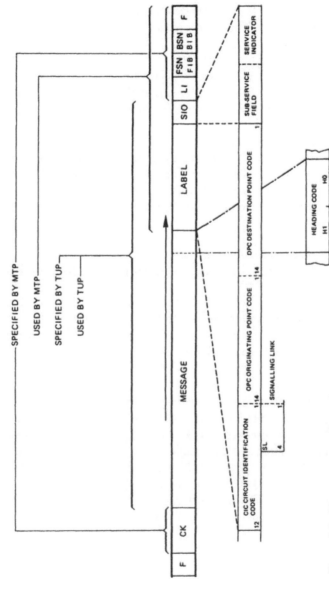

Figure 4.7 The system no. 7 envelope.

the address information once for the MTP and once for the user part, provided the address information used a common format. This, sadly, is not the case.

4.1.4 Introduction to #7 user parts

Figure 4.5 indicates the existence of the user parts, or entities that can be considered as user parts. In the figure these are listed in the order in which they are treated by the CCITT recommendations, but they are listed below in their order of appearance as defined protocols.

4.1.4.1 Telephone user part (TUP)
This defines the protocols necessary to set up, clear down and charge telephone calls on international circuits. It is analogous, therefore, to CCITT R2 or no. 5 and embraces much the same functionality[50-54].

4.1.4.2 Data user part (DUP)
This defines the protocols necessary to set up, clear down and charge data calls circuit switched between users. The user entities may well be administration functions belonging to the network providers. The protocols include provision for sub-64kbit/s connections. (Recommendation Q.741[55] refers to X.61[56])

4.1.4.3 ISDN user part (ISDN UP)
This defines the additional protocols necessary on an ISDN call over and above those defined by TUP[57-61].

4.1.4.4 Signalling connection control part (SCCP)
This defines the additional requirements for non-circuit-related signalling. As such, the SCCP is really an adjunct of the MTP and, with the MTP, forms the network service part (NSP) (Fig. 4.8)[62-66].

4.1.4.5 Digital subscriber signalling system no. 1 (DSS 1)
It is in these recommendations that CCITT gives official recognition to administration resistance to user access to #7 signalling. DSS 1 recommendations are the most closely aligned to OSI principles and specify link access procedure on the D channel (LAPD) based upon LAPB (Ch. 2) and HDLC (Ch. 2). Recommendations Q.930[35] and Q.931[36] specify the OSI layer 3 functions.

As the fundamental means of providing ISDN access, DSS 1[33-36] is returned to and treated in more detail later in this chapter.

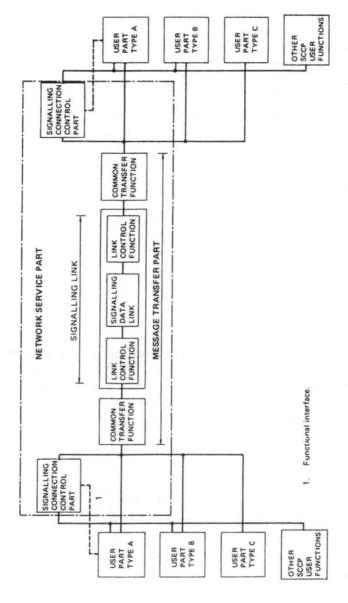

Figure 4.8 Function of the SCCP[60]. The ISDN UP is defined in Q.761–Q.764[57–60], Q.766[61] and Q.767[67].

4.1.5 #7 user parts, structure

The remainder of this treatment of the user parts will concentrate upon the differences in approach between the most important parts, that, even from the brief details given above, emerge as the TUP, DUP and ISDN UP.

Because of the differences in user part protocols for address information that we are about to discuss it is necessary for the MTP to know to which user part the message belongs. This information is provided to the MTP by the service information octet SIO as illustrated in Figure 4.9.

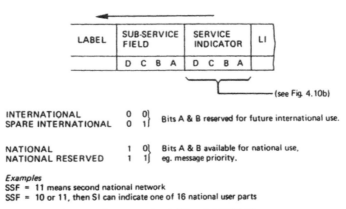

Figure 4.9 Service information octet (SIO) (Fig. Q.704/13)).

4.1.5.1 #7 user parts: the label

The first bits of the #7 message constitute the label. Figure 4.7 illustrates this and shows in detail the label used by both TUP and ISDN UP. That this is not the end of the story is illustrated by Figure 4.10, which contrasts the labels of MTP, TUP and ISDN UP, and DUP. All three agree in providing address data for originating and destination points. The signalling link selection (SLS) portion of the MTP label is inherent in the other labels as it is the least significant digits

	SLS	OPC ORIGINATING POINT	DPC DESTINATION POINT CODE	MTP Q.704
	4 1 14		1 14 1	

SLS = SIGNALLING LINK SELECTION (see *Note*)

	CIC CIRCUIT IDENTIFICATION CODE	OPC	DPC	TUP Q.723 / ISDN UP / Q.763
	12 1			

TSC TIME SLOT CODE	BIC BEARER IDENTIFICATION CODE	OPC	DPC	DUP X.61
8 1	12 1			

Figure 4.10 Label definitions. Note that Recommendation Q.704, para 14.2, replaces SLS with signalling link code (SLC) for signalling network management messages.

of the circuit identification or bearer identification codes. The SLS is needed to control load-sharing functions to ensure that related messages are not so routed that they arrive at their destination in the wrong order.

The remaining differences in the label between TUP and DUP are related to the fact that the TUP always deals with 64 kbit/s channels whereas the DUP has facilities for submultiplexing. In the DUP label the bearer identification code (BIC) identifies down to 64 kbit/s and the time slot code (TSC) identifies below this.

4.1.5.2 #7 user parts: the message
The MTP requires a repertoire of network control messages that can be defined in complete detail. To a lesser degree of definition the TUP messages are definable. The DUP message, by contrast, is less well defined.

There are difficulties inherent in attempting to define a universal standard for diverse services. An example is the need for the MTP to delve even deeper into the message to distinguish between, say, an MTP changeover message and a DUP message with the same coding for H0 and H1.

By the time that the CCITT study group (SG XVIII) came to write the ISDN UP recommendations a much improved method of presentation had been devised, and this is evident in Figure 4.11, which shows the ISDN UP messages. Much the same procedure for mandatory and optional portions is being used as in the DUP with the additional complication of fixed-length and variable-length messages. The method is explained by showing a full message format in Figure 4.12 with explanatory notes and references to illustrate how it is constructed. Note that the mandatory parameters are expected because of the message type and are not therefore named. Note, too, that the length indicator gives the parameter length in octets, including the parameter name (if optional) and the length indicator itself.

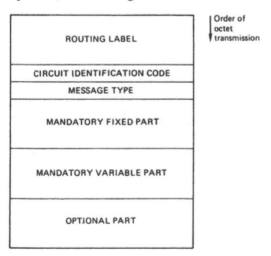

Figure 4.11 ISDN user part, message parts[58].

(a)

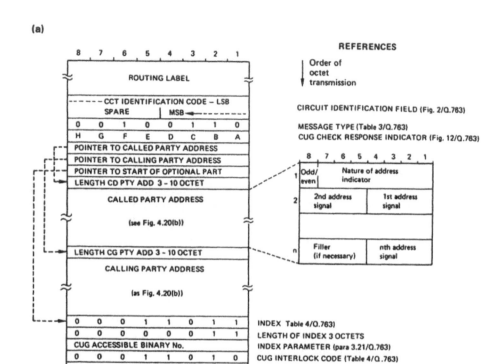

(b)

Parameter	Reference (para)	Type	Length (octets)
Message type	2.1	F	1
Closed user group check response indicators	3.12	f	1
Called party address	3.6	V	3 to 10
Calling party address	3.7	V	3 to 10
Index	3.21	O	3
Closed user group interlock code	3.13	O	6
Call reference	3.5	O	7

Figure 4.12 (a) Closed user group selection and validation response message and called party address parameter field[58]. (b) Message type: closed user group selection and validation response[58].

90

4.1.5.3 User parts review

What has been said indicates that the term "user part" has become a misnomer. Only DSS 1 (Recommendations Q.920–Q.931) refers to signalling facilities that are open to use by the ISDN user. TUP, DUP and ISDN UP all refer to signalling features used by the network provider on behalf of the user. In touching upon the users' specific interests in Q.920 and Q931 we have come closest to using OSI concepts, which still await treatment in the second half of this chapter.

The difficulties that have appeared and complicated #7 as a message transport mechanism because of the different objectives of the defining study groups responsible for the different user parts have been outlined. A signalling concept that was conceived to have universal application has become complex in execution. The conceptual division into parts, for transport and for disparate users, has, in practice, been eroded.

To complete a rather gloomy picture it is now necessary to consider briefly the CCS system devised specifically for the user.

4.1.6 CCS for users

It was perhaps BT, then the only UK national network provider, who invented the "dog in a manger" attitude to CCS: "the user shall not have access to the administration's signalling system". The reasons have been touched upon already and stem, in part, from the antics of certain Americans with "blue boxes" when subscriber touch tone signalling was first introduced in the US.[3]

BT was also the first administration to make a commercial offering of ISDN facilities and did so before the ISDN access arrangements were finalized. The BT offer, known as integrated digital access (IDA), therefore went into operation using BT's original access proposal of 2B+D channels in 80 kbit/s, namely B 1 (64 kbit/s), B2 (8 kbit/s) and D (8 kbit/s), not the basic access eventually defined (illustrated in Fig. 3.2).

As with the development of #7, BT therefore forged ahead of international agreement and defined for the IDA a CCS for subscribers use known as DASS 1. Before the IDA was launched the UK PABX manufacturers and BT co-operated in the development of a CCS for private use on digital private circuits between PABX units. This system was conceived to enable PABX facilities to be extended over the private network, allowing the complete private network to behave as if it were a single PABX. To an extent this was realized by the definition of digital private network signalling system (DPNSS) based upon DASS 1.

The DPNSS development identified deficiencies in the original DASS 1 specification in that DPNSS was a symmetrical system whereas DASS 1 was access

---

3. These "phone freaks" were, paradoxically, in part responsible for the development of the personal computer which, as we have seen, has added urgency to ISDN development [68].

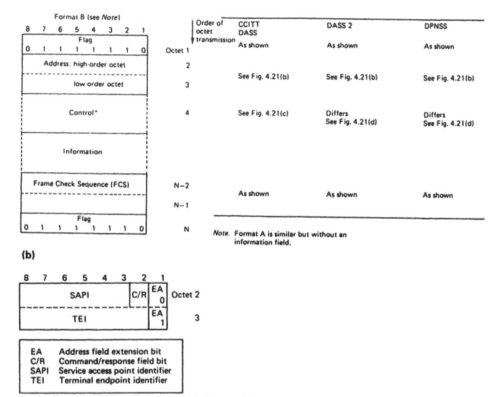

(a)

		Order of octet transmission	CCITT DASS	DASS 2	DPNSS
	Octet 1		As shown	As shown	As shown
	2				
	3		See Fig. 4.21(b)	See Fig. 4.21(b)	See Fig. 4.21(b)
	4		See Fig. 4.21(c)	Differs See Fig. 4.21(d)	Differs See Fig. 4.21(d)
	N−2				
	N−1		As shown	As shown	As shown
	N				

Note. Format A is similar but without an information field.

(b)

EA Address field extension bit
C/R Command/response field bit
SAPI Service access point identifier
TEI Terminal endpoint identifier

Note Single octet address field is reserved for LAPB operation

8	7	6	5	4	3	2	1	VALID ADDRESS FIELD COMBINATIONS	
								DASS2	DPNSS
0	0	0	0	1	1	0	0	SABMR from PBX	SABMR from PBX B
0	0	0	0	1	1	1	0	UA from PBX	UA from PBX B
0	0	0	0	0	1	0	0	UI(C) from PBX or UI(R) from ET	———
0	1	0	0	0	1	0	0	———	UI(C) from PBX B or UI(R) from PBX A
0	0	0	0	1	1	1	0	SABMR from ET	SABMR from PBX A
0	0	0	0	1	1	0	0	UA from ET	UA from PBX A
0	0	0	0	0	1	1	0	UI(C) from ET or UI(R) from PBX	———
0	1	0	0	0	1	1	0	———	UI(C) from PBX B or UI(R) from PBX A

Figure 4.13 (a) User CCS LAPD frame formats. (b) User CCS address field[34].

(c)

Command	Response	8	7	6	5	4	3	2	1	Octet
					Encoding					
Information frames										
I Frame				N(S)					0	4
				N(R)					P	5
Supervisory frames										
RR	RR	0	0	0	0	0	0	0	1	4
					N(R)				P/F	5
RNR	RNR	0	0	0	0	0	1	0	1	4
					N(R)				P/F	5
REJ	REJ	0	0	0	0	1	0	0	1	4
					N(R)				P/F	5
Unnumbered frames										
SABME	-----	0	1	1	P	1	1	1	1	4
-----	DM	0	0	0	F	1	1	1	1	4
UI	-----	0	0	0	P	0	0	1	1	4
DISC	-----	0	1	0	P	0	0	1	1	4
-----	UA	0	1	1	F	0	0	1	1	4
-----	FRMR	1	0	0	F	0	1	1	1	4

I Frame	Information frame	DISC	Disconnect
RR	Receiver ready	UA	Unnumbered acknowledge
RNR	Receiver not ready		
REJ	Reject	FRMR	Frame reject
SABME	Set asynchronous balanced mode extended	DM	Disconnect mode
		UI	Unnumbered information
N(S)	Transmitter send sequence number	N(R)	Transmitter receive sequence number
S	Supervisory function bit		
P/F	Poll bit when issued as a command, final bit when issued as a response	M	Modifier function bit
		X	Reserved and set at 0

(d)

CONTROL FIELD BITS (MODULO 8)	8	7	6	5	4	3	2	1
UI frame	0	0	0	N	0	0	1	1
SABMR	1	1	1	0	1	1	1	1
UA frame	0	1	1	0	0	0	1	1

N = sequence number of UI frame

Figure 4.13 (c) CCITT DSS 1 control field[34]. (d) DASS 2 and DPNSS control field.

Table 4.1 DSS 1 protocol specifications.

Layer 1	I.430	[22]	Basic layer 1 specification
	I.431	[23]	Primary rate layer 1 specification
Layer 2	Q.920	[33]	Data link layer, general aspects
	Q.921	[34]	Data link layer, specification
Layer 3	Q.930	[35]	Layer 3, general aspects
	Q.931	[36]	Layer 3, specification for basic call control
	Q.932	[37]	Generic procedures for control of ISDN supplementary services
	Q.933	[70]	Layer 3, specification for frame-mode bearer services
	Q.950	[71]	ISDN supplementary services
	Q.951 to Q.957	[72–77]	Stage 3 descriptions for ISDN supplementary services

asymmetric, and DASS 2 was developed to rectify this, thereby allowing DPNSS transport via DASS 2 over private digital circuits in the public network. DASS 2 further allows for the ISDN access arrangement defined by Recommendations I.430 and I.431. The primary access supported is, of course, the 2.048 Mbit/s version of I.431.

We have, in consequence, at least three CCS defined for user access: CCITT DSS 1 (Recommendations Q.920–Q931), DASS 2 for PABX to network use and DPNSS for PABX to PABX use via, if necessary, DASS 2 and the network. The similarities and differences between these three extant, defined systems are illustrated in Figure 4.13[69].

In other national networks the gaps in early definitions of CCITT DSS 1 were completed by some form of manufacturer-specific or network-specific signalling system. All these individual systems will eventually be replaced by CCITT DSS 1, but as ISDN applications have proliferated the task of standardization has become more difficult.

One good reason for including the LAPD formats depicted in Figure 4.13 is to highlight the departure in user CCS from the error control mechanism inherent in the #7 MTP. The reasons for this are historical but none the less questionable. At the time that the UK administration and others were considering ISDN trials and devising suitable user CCS, the UK and Europe were introducing packet switching. The packet-switched data network (PSDN) uses HDLC and LAPB, as was described in Chapter 2, and the question of whether to harmonize with an existing (although small) PSDN or with the network CCS (#7) when integration of user CCS is generally considered a "bad thing" was the subject of heated debate. Inevitably, the PSDN won and a very comprehensive error control protocol (the #7 MTP) was discarded in favour of a rather rudimentary protocol capable of little more than compelled retransmission. The only thing that both protocols have in common is the same error check algorithm.

4.1.6.1 CCITT user–network access signalling: DSS 1

So far we have discussed ISDN access by telling the story of how it has come to be as it is. Much of this story is, in the author's view, important in illustrating how technology can develop in directions that are not always optimum; it is also my belief that, by recognizing this, future developments will be better managed to achieve optimum results. The story too has demonstrated how technology is influenced by many factors, from government policy to the interests of network operators, manufacturers and, possibly, the users. For this second edition, which describes ISDN as a fact rather than a possibility, it is necessary to recapitulate the material on ISDN access, the fundamental technology that makes ISDN possible. This recapitulation will necessarily refer to the areas concerning ISDN access already covered in the story.

The user–network interface, as we have seen, has been defined in terms of the physical arrangements (Figs 3.3 and 3.6) and the signalling protocols applicable at layers 1, 2 and 3 of the OSI model. Chapter 3 has dealt with the physical interface characteristics defined at layer 1, and this section will deal with the protocols at each of the three layers. The CCITT recommendations defining each of the three layers of the DSS 1 protocol are shown in Table 4.1.

Layer 1 (basic user–network interface) is responsible for transferring information between terminals and the NT1 (Fig. 3.6). The functions of layer 1 are as follows:

(a) B channels: layer 1 must support, for each direction of transmission, two independent 64 kbit/s B channels.

(b) D channels: layer 1 must support, for each direction of transmission, a 16 kbit/s D channel for signalling information and for user packet data if this is supported.

The structures of the layer 1 frame that permit support of these communications channel requirements across the interface are different in both directions of transmission; they are shown in Figure 3.9 and are described in the accompanying text.

D channel access procedure is a procedure to ensure that when two or more terminals on the passive bus (Fig. 3.3) attempt to access the D channel simultaneously, one terminal only will be successful. This access procedure, using CSMA-CD is described in Chapter 3.

D-channel Deactivation permits the TE and NT equipments to be placed in a low-power mode when no calls are in progress. Activation may be initiated from either the terminal or the network; deactivation can only be initiated by the network because only the network knows that all the terminals on the ISDN basic access are quiescent.

D-channel activation and deactivation depend on the use of four types of signal between the terminal and the network:

(a) INFO 0: no signal, i.e. neither the terminals nor the NT1 are working. The TE is in its low-power, dormant state.

(b) INFO 1: TE to NT only. Transmitted by any terminal when first activated. Consists of a particular pattern; +0, –0, 1, 1, 1, 1, 1, 1. This is, of course, not synchronized to the network and, if more than one terminal activates at the same time, may consist of a mixture of several such signals. On receipt and recognition of INFO 1 the NT will respond with:

(c) INFO 2: NT to TE only. A particular, defined, frame with a high density of binary zeros in the data channels to permit the terminals to synchronize quickly to the signals. When synchronized the TE can respond with:

(d) INFO 3: TE to NT. Synchronous frames containing operational data. On receipt of INFO 3 frames the NT can respond with:

(e) INFO 4: NT to TE. Synchronous frames containing operational data.

The terminals are now "woken up" and communication has commenced and can proceed. Note that all the terminals on the passive bus are activated at once: it is not possible to have only one terminal active in a multiterminal configuration.

The differences between the primary rate interface and the basic interface are, to a great extent, relevant to layer 1 of the interface only.

(a) No provision is made in the primary rate interface for multipoint work-ing – the connection will be between the network and a single device (a PABX 30-channel exchange line interface, for example). Individual chan-nels across the interface may, however, be routed independently as required.

(b) No provision is made in the primary rate interface for deactivating the link to economize on power feeding. In general, power feeding does not take place across the primary rate interface.

(c) Customers using the primary rate access can be expected to use indi-vidual channels for a variety of purposes including private lines as well as channels to the public telecommunications network. It was to make pro-vision for this in advance of complete definition by the CCITT that the UK solution of using DPNSS (private network signalling) and DASS 2 (public, user–network signalling) was introduced.

4.1.6.2 Layer 2 access protocol

The objective of layer 2 is to provide a secure, error-free connection between two endpoints connected by a physical medium. Layer 3 call control informa-tion is carried in the information elements of the layer 2 frames and must be delivered in sequence and without error. Layer 2 must also detect and retransmit lost frames.

As we have said, LAPD was based on LAPB defined for layer 2 of the X.25 protocol. Certain features of LAPD give it significant advantages, including the possibility of permitting frame multiplexing by having separate addresses at layer 2, allowing many LAPs to exist on the same physical connection. This feature makes it possible for up to eight terminals to share the signalling chan-nel in the passive bus arrangement.

Each layer 2 connection is a separate LAP and the termination points for the LAPs are within the user terminals at one end and at the periphery of the exchange at the other. Layer 2 acts as a series of frame exchanges between the two communicating (or peer) entities. The frames consist of a series of eight-bit elements and their position in the sequence defines their meaning, as shown in Figure 4.13a. A fixed pattern flag defines both the beginning and end of the frame. Two octets carry the layer 2 address in the form of a service identifier (SAPI; see below), a terminal identifier (TEI; see below) and a command/response bit (see Fig. 4.13b). The extension bit distinguishes the LAPD, dual-octet address from, for example, the LAPB single-octet address (Fig. 2.16).

The control field may be one or two octets in length and carries informa-tion to identify the frame and the layer 2 sequence numbers used for link con-trol (Fig. 4.13c). The information field is only present in frames that carry layer 3 information and the frame check sequence (FCS) is used to detect errors. More details are contained in Figure 4.13, which also explains the abbreviations.

4.1.6.3 Layer 2 addressing

Multiplexing of layer 2 messages is achieved by using a separate layer 2 address for each LAP in the system. The two-octet address field identifies the intended receiver of a command frame and the transmitter of a response frame. Thus the address has only local significance and is known and used only by the two endpoints using the LAP.

The construction of the layer 2 address is shown in Figure 4.13b. The service access point identifier (SAPI) is used to identify the service that the signalling frame is intended to use. At present, only signalling and packet-mode services have been identified.

Apart from the case of the global, or broadcast, TEI (TEI = 127) the terminal endpoint identifier (TEI) has a unique value associated with a single terminal on the customer access. Values of TEI = 0–63 are allocated by the user equipment, and it is the user's responsibility to ensure that they are unique. Values of TEI = 64–126 are allocated by the network when the user invokes the automatic TEI assignment procedure and the network ensures that a unique value is allocated.

Illustrating this arrangement, a frame originating from telephony call control has a SAPI that identifies the frame as "telephony" (SAPI = 0), and all telephone equipment will examine this frame. Only the terminal whose TEI matches that carried by the frame will pass it to the layer 2 and layer 3 entities for processing.

The global TEI is used to broadcast messages to all terminals within a given SAPI. This could be, for example, a broadcast message to all telephones offering an incoming telephony call.

4.1.6.4 Layer 2 operation

The procedure for setting up a call that has been described is summarized in Figure 4.14.

4.2 Interoperability

So far we have discussed the terminals that are interconnected and the means by which this interconnection is effected. We have not discussed in detail how the terminals will interoperate, i.e. co-operate in treating and understanding the data they share between them. The discussion is mainly limited to the OSI model, but one other protocol deserves mention because of its widespread application in data processing interoperability. This is Transmission Control Protocol/Internet Protocol (TCP/IP). This protocol was not discussed in the first edition even though at that time TCP/IP was more widely deployed than any OSI-based protocol because it has only been the emergence of the B-ISDN with its ability to interconnect LANs with high-capacity data links that has required a degree of understanding of TCP/IP. This will be discussed in Section 4.2.2.

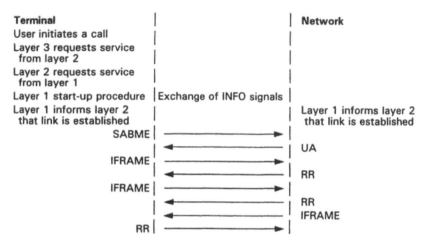

Figure 4.14   Call initiation by a terminal using layers 1 and 2.

4.2.1 Open systems interconnection (OSI)

Chapter 2 introduced the problems of communicating between different machines and the consequent need for protocol. The discussion was taken as far as the means to transport data from machine to machine over a network. The subsequent discussion of CCS provided more detail on the available transport mechanisms more specifically for telecommunications networks, of which the ISDN is one example. Very little of this discussion helped to explain how the recipient machine can understand the data presented by the transport mechanism.

While the telecommunications network providers were puzzling over the problems of communicating communications processors and solving these problems with CCS, the computer manufacturers and users puzzled over the analogous problem of communicating computers. In the computer industry there was more rivalry and more urgency. The problem appeared to simplify at times into the (computer) world seeking to communicate with IBM. The non-IBM world invoked the ISO and the OSI model was the result. IBM, on the other hand, introduced a similar model, systems network architecture (SNA), and the resolution of OSI and SNA into a single, universal model took many years to achieve.

Where the idea of layering originated is not clear, at least to the author, but it is fundamental to CCS, particularly #7, and to OSI and SNA. As will be seen, however, the #7 layers are not entirely analogous to some of the OSI layers but neither is there undue conflict.

When the telecommunications industry addressed the problem of the ISDN the OSI model was well established and it was perhaps natural, certainly right, to organize the ISDN in accordance with the OSI model. The ISO definition of the OSI model is contained in ISO 7498/DAD1[78]. However, CCITT has republished the definition without change but with substantial introductory material applicable to telecommunications applications in CCITT Recommendation X.200[79]. Apart from the introductory matter Recommendation X.200 is an exact copy of ISO 7498 with very few explanatory additions.

4.2.2 The OSI model

The OSI model was briefly introduced in Chapter 1 (Fig. 1.7 is relevant) and has been constantly referred to since. Now, a formal introduction will be provided by reprinting the explanation of how the layers were chosen from Annex A of ISO 7498. To appreciate this narrative, however, it is first necessary to list the principles upon which the OSI model is based.

4.2.2.1 OSI: Principles of layering
The following passage is copied directly from Section 6.2 of X.200. ISO 7498 differs only in using the term "international standard" instead of "recommendation".

6.2 The principles used to determine the seven layers in the Reference Model

The following principles have been used to determine the seven layers in the reference model and are felt to be useful for guiding further decisions in the development of *OSI* recommendations.

Note.   It may be difficult to prove that any particular layering selected is the best possible solution. However, there are general principles which can be applied to the question of where a boundary should be placed and how many boundaries should be placed.

P1:   do not create so many layers as to make the system engineering task of describing and integrating the layers more difficult than necessary;

P2:   create a boundary at a point where the description of services can be small and the number of interactions across the boundary are minimized;

P3:   create separate layers to handle functions that are manifestly different in the process performed or the involved technology;

P4:   collect similar functions into the same layer;

P5:   select boundaries at a point which past experience has demonstrated to be successful;

P6:   create a layer of easily localized functions so that the layer could be totally redesigned and its protocols changed in a major way to take advantage of new advances in architectural, hardware or software technology without changing the expected services from and provided to the adjacent layers;

P7: create a boundary where it may be useful at some point in time to have the corresponding interface standardized;

Note 1 Advantages and drawbacks of standardizing internal interfaces within open systems are not considered in this recommendation. In particular, mention of, or reference to, principle P7 should not be taken to imply usefulness of standards for such internal interfaces.

Note 2 It is important to note that *OSI per se* does not require interfaces within open systems to be standardized. Moreover, whenever standards for such interfaces are defined, adherence to such internal interface standards can in no way be considered as a condition of openness.

P8: create a layer where there is a need for a different level of abstraction in the handling of data, e.g. morphology, syntax, semantics;

P9: allow changes of functions or protocols to be made within a layer without affecting other layers; and

P10: create for each layer boundaries with its upper and lower layer only.

Similar principles have been applied to sublayering:

P11: create further subgrouping and organization of functions to form sublayers within a layer in cases where distinct communications services need it;

P12: create, where needed, two or more sublayers with a common, and therefore minimal functionality to allow interface operation with adjacent layers; and

P13: allow bypassing of sublayers.

These principles, laudable in intent, do contain much that is dictated by the real world. P5, for example, justifies retention and utilization of existing layered structures (X.25) and existing interface protocols, V.24, HDLC, etc. This is not a criticism but a caution; the ten principles were not brought down from the technical equivalent of Mount Sinai.

4.2.2.2 OSI: how the layers were chosen

With this background it is possible to enjoy the narrative of Annex A to both X.200 and ISO 7498. It is a pity perhaps that both documents are so structured as to keep this, the most enlightening passage, until the very end.

This annex provides elements giving additional information to this recommendation, which are not an integral part of it.

The following is a brief explanation of how the layers were chosen:

(a) It is essential that the architecture permit usage of a realistic variety of physical media for interconnection with different control procedures (e.g. V.24, V.25, etc.). Application of principles P3, P5 and P8 leads to identification of a physical layer as the lowest layer in the architecture.

(b) Some physical communication media (e.g. telephone line) require specific techniques to be used in order to transmit data between systems despite a relatively high error rate (i.e. an error rate not acceptable for the great majority of applications). These specific techniques are used in data link control procedures which have been studied and standardized for a number of years. It must also be recognized that new physical communication media (e.g. fibre optics) will require different data link control procedures. Application of principles P3, P5 and P8 leads to identification of a data link layer on top of the physical layer in the architecture.

(c) In the open system architecture, some open systems will act as the final destination of data (see division 4). Some open systems may act only as intermediate nodes (forwarding data to other open systems) (see Figure 13/X.200). Application of principles P3, P5 and P7 leads to identification of a network layer on top of the data link layer. Network-oriented protocols such as routing, for example, will be grouped in this layer. Thus, the network layer will provide a connection path (network connection) between a pair of transport entities, including the case where intermediate nodes are involved (see Figure 13/X.200) (see also §7.5.4.1).

(d) Control data transportation from source end open system to destination end open system (which is not performed in intermediate nodes) is the last function to be performed in order to provide the totality of the transport service. Thus, the upper layer in the transport service part of the architecture is the transport layer, on top of the network layer. This transport layer relieves higher layer entities from any concern with the transportation of data between them.

(e) There is a need to organize and synchronize dialogue, and to manage the exchange of data. Application of principles P3 and P4 leads to the identification of a session layer on top of the transport layer.

(f) The remaining set of general interest functions are those related to representation and manipulation of structured data for the benefit of application programs. Application of principles P3 and P4 leads to identification of a presentation layer on top of the session layer.

(g) Finally, there are applications consisting of application processes which perform information processing. An aspect of these application processes and the protocols by which they communicate comprise the application layer as the highest layer of the architecture.

The resulting architecture with seven layers, illustrated in Figure 12/X.200, obeys principles P1 and P2.

A more detailed definition of each of the seven layers identified above is given in division 7 of this recommendation, starting from the top with the application layer described in section 7.1 down to the physical layer described in section 7.7.

Figure 12/X.200 is reproduced here as Figure 4.15, and Figure 13/X.200 as Figure 4.16.

It is perhaps mischievous to note that the explanations become briefer as we reach the higher levels of abstraction and these are the layers that are less well defined. Both transport and session layers are introduced without any principles at all, and principles 6, 9 and 10 are not needed to justify any of the

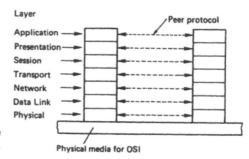

Figure 4.15 Seven-layer reference model and peer protocols[79].

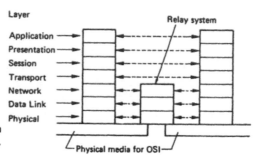

Figure 4.16 Communication involving relay open systems[79].

layers. Principle 1, at least, could be invoked to question the existence of as many as seven layers.

4.2.2.3 OSI: the concept

The purpose and concept of the OSI model was to develop a model that would allow the connection of real systems and to allow their interworking, provided they complied with the model, to be "open". The model does not itself define protocols or standards but provides an abstract model upon which the necessary standards may be based. Interconnection implies a physical interconnecting medium (Fig. 4.17), but the kind of medium envisaged can be quite varied. The OSI model could, for example, assist the physical transfer of material (by carrying disks, tapes, etc.) between systems. In such circumstances, perhaps, the lower layers of the model might be redundant.

Figure 4.17 Open systems connected by physical media[79].

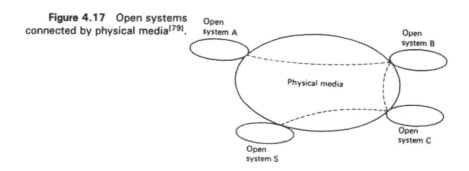

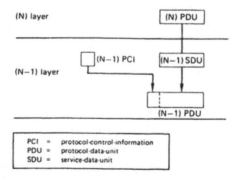

Note 1. This figure assumes that neither segmenting nor blocking of (N)-service-data-units is performed (see para 5.7.6.5).

Note 2. This figure does not imply any positional relationship between protocol-control-information and user-data in protocol-data-units.

Note 3. An (N)-protocol-data-unit may be mapped one-to-one into an (N−1)-service-data-unit, but other relationships are possible (see Figure 11/X.200).

Figure 4.18 An illustration of mapping between data units in adjacent layers[79].

Central to the OSI concept is the idea of layers. Each layer can only communicate with its peer(s) and does so by using a service offered to it by the layer below (Figs 4.15 and 4.16). Each layer therefore can be fully described by: *purpose*; *service* offered to the layer above; *functions* within the layer; *service* used from the layer below. Such a concept does not inherently demand the envelope technique of the layer-to-layer protocol described in Chapter 2 (Fig. 2.15 for example), but it is difficult to imagine any other kind of protocol that would meet the requirements of the model. The model description does indeed assume enveloping, although it does not demand it, and Figure 4.18 illustrates the presence of this assumption.

4.2.2.4 OSI: applications

Having introduced OSI, it is now appropriate to return to subjects already treated and see the application of OSI in practical situations.

(a) OSI in packet switching. In Chapter 2 we developed an argument explaining the successive requirements for secure, meaningful communication. This culminated in Figure 2.11, which, in fact, defined the protocols of packet switching and of ISDN. It is easy to see now that this also showed a particular implementation of the lower four layers of the OSI model, and Figure 4.19 reproduces the same information as Figure 2.11 in a way that emphasizes this. Messages entering and leaving layer 4 are communications independent, and only layer 4 has any intercourse with the higher layers that use and generate the messages. However, nothing in the functions provided by layers 1–4 helps the communicating machines or processors to understand the content of the message.

(b) OSI in CCITT digital subscriber signalling system no. 1 DSS 1 (Q.920–931). The data link protocol LAPD adopted for ISDN is a variation on

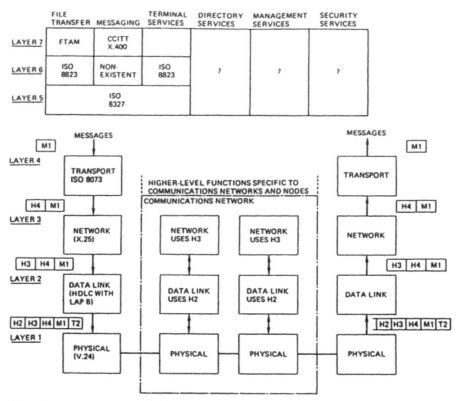

	FILE TRANSFER	MESSAGING	TERMINAL SERVICES	DIRECTORY SERVICES	MANAGEMENT SERVICES	SECURITY SERVICES
LAYER 7	FTAM	CCITT X.400				
LAYER 6	ISO 8823	NON-EXISTENT	ISO 8823	?	?	?
LAYER 5	ISO 8327					

Note Bracketed protocols are typical candidates used in text.

Figure 4.19 OSI application in PSS. ISO 8073, connection-oriented transport; ISO 8327, connection-oriented session; ISO 8823, connection-oriented presentation; CCITT X.400, message-handling systems. FFTAM, file transfer, access and management.

LAPB allowing more than one communication over the same physical channel. The LAPD format was illustrated in Figure 4.13a. The effect of LAPD is illustrated in Figure 4.20, which contrasts a direct user interface with the PSDN with a similar interface via ISDN. Figure 4.21 illustrates the use of the LAPD SAPI and TEI addresses to distinguish between several user processes on the same physical channel.

An American interpretation of DSSI in OSI terms is illustrated in Figure 4.22. This goes further in suggesting protocols for the higher layers for use in a specific task, that of file transfer using X.25 packet handling over the D channel.

(c) OSI in CCITT #7 signalling. An important example is to attempt to reconcile the #7 signalling system with the OSI model. The user part of #7 is so called with no intention of implying availability to the user as a per-

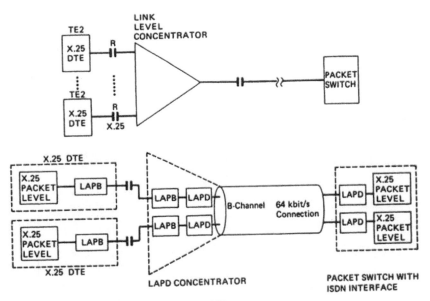

Figure 4.20 Use of LAPD on a B channel[80].

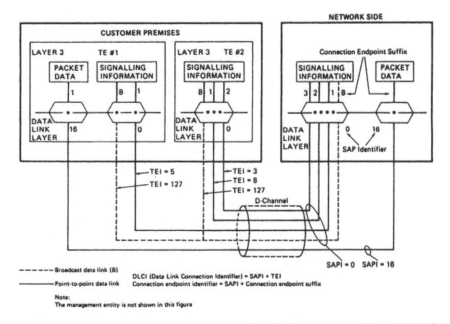

Figure 4.21 Overview description of the relationship between SAPI, TEI, and data link connection endpoint identifier[80].

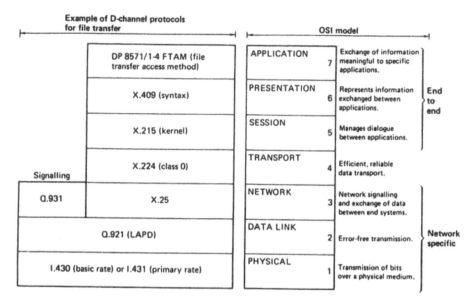

Figure 4.22 The OSI reference model and D channel protocols[82].

son. The TUP for instance provides the features for the user systems, i.e. telephone exchanges, to provide connection services using the signalling system. #7, although acknowledged to be a powerful communications medium, is defined as an administration tool. With this proviso then, Figure 4.23 is offered as an attempt to reconcile #7 with the OSI model.

4.2.3 Transmission control protocol/Internet protocol (TCP/IP)[4]

TCP/IP was developed in the 1970s as part of the development of Arpanet, a wide area network linking research centres of the US Government Defense Advanced Research Projects Agency. Thus, it precedes the OSI model by several years, but it uses a layered architecture that fits quite well with the OSI model. When the University of California at Berkeley distributed the Unix operating system it included TCP/IP in the package and this, in part, explains its popularity in the computer industry.

TCP/IP defines no specific protocols for OSI layers 1 and 2. The Internet protocol is the TCP/IP layer 3 protocol. It is used successfully with Ethernet, StarLAN and in conjunction with X.25-compatible layer 1 and 2 protocols. The Internet protocol provides host-to-host delivery of data across the network. It does not concern itself with the order in which data arrives or with the reliability of the transmission, only with the routing.

---

4. For much of this discussion I am, not for the first time, indebted to an excellent little book by Richard Bowker[83].

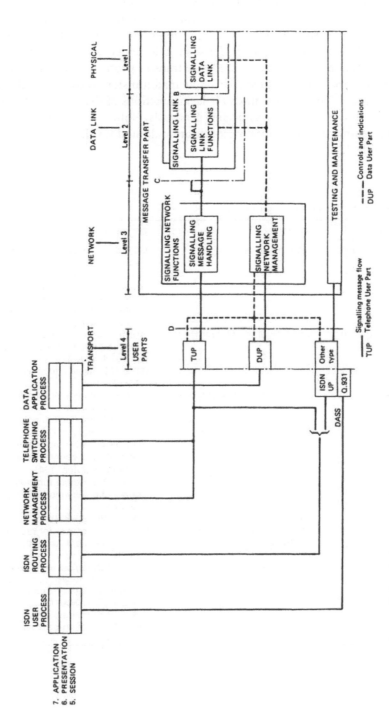

Figure 4.23 System no. 7 signalling as an OSI application (Note: for an alternative view, see Figure 3 of ref. 82).

The transmission control protocol operates at the equivalent of layer 4 of the OSI model. It provides a "virtual circuit" connection between the transmitting and receiving entities, making it appear as if there is a permanent connection. It also provides flow control, ensuring that a slower receiving device is not swamped with data. It also ensures that data is received reliably, that it is in the correct sequence, that none is missing and that none is duplicated.

Associated with TCP/IP are several application utilities that fit into the higher layers of the OSI model. These include Telnet, providing remote log in and virtual terminal support; FTP (file transfer protocol), which permits the user to transfer files; and SMTP (simple mail transfer protocol), an electronic mail utility. Table 4.2 contrasts the layer functions of the OSI model and TCP/IP and shows some of the associated protocols. Note that this is not an exhaustive treatment: many more protocols are indicated for the lower three layers in Table 4.1.

Table 4.2 OSI and TCP/IP protocols.

Layer	TCP-IP	OSI model
Application	File transfer protocol	ISO 8650: common application service elements ISO 8571: FTAM ISO 9041: virtual terminal ITU-T X.400: message handling ISO 8650: association control service elements
Presentation	Simple mail transfer protocol	ISO 882: connection-oriented presentation protocol ISO 8824: ASN 1 ISO 8825: basic encoding rules for ASN 1
Session	Telnet virtual terminal protocol	ISO 8327: connection-oriented session protocol
Transport	Transmission control protocol User datagram protocol	ISO 8073: connection oriented transport protocol ISO 8073, Add 2: operation of class 4 over connectionless network service (TP4) ISO 8602: protocol for providing connectionless-mode transport service
Network	Internet protocol Ethernet address resolution protocol	ITU-T X.25, ISO 8208 ISO 8373: protocol for providing the connectionless-mode network service
Data link	IEEE 802 protocols ITU-T X.25	IEEE 802.2: logical link control IEEE medium access control and high-level data link control (HDLC)
Physical		IEEE 802.3: CSMA/CD IEEE 802.4: token-passing bus IEEE 802.5: token-passing ring ITU-T X.21

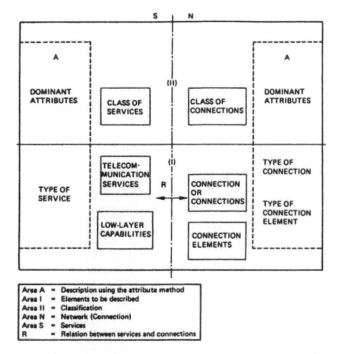

Figure 4.24 Concepts used to characterize telecommunications services and connections[84].

4.3 Characterizations of services and network capabilities

A descriptive method, present inherently in earlier CCITT documents, has become formalized and is used extensively in the I-series recommendations on the ISDN. It is a notable omission that this book has progressed so far without using the term *attribute*. It would be wrong to leave the topic of ISDN tools without providing details of a method that has been a significant "tool" in constructing the ISDN recommendations. An attribute is a specific characteristic of an object or element (CCITT terms, "type of service", "type of connection", are objects or elements) whose values distinguish the object or element from others. To describe an ISDN service or network capability, therefore, it is necessary to list its attributes and assign values to the attributes. Figure 4.24 illustrates the method. Figure 4.25 tabulates the possible attributes of bearer services supported by the ISDN.

Chapter summary

If the saying that one picture is worth a thousand words is true then this is the best chapter in the book, relying, as it does, on its many figures to convey

Possible values of attributes							Attributes (Note 6)
							Information transfer attributes
Circuit					Packet		1. Information transfer mode
Bit rate (kbit/s)					Throughput		
64	384	1536	1920	Other values for further study	Options for further study		2. Information transfer rate
Unrestricted digital information	Speech	3.1kHz audio	7kHz audio	15kHz audio	Video	Others for further study	3. Information transfer capability
8kHz integrity		Service data unit integrity (Note 2)		Unstructured			4. Structure
Demand		Reserved		Permanent			5. Establishment of communication (Note 5)
Point-to-point		Multipoint		Broadcast (Note 1)			6. Communication configuration
Unidirectional		Bidirectional symmetric		Bidirectional asymmetric			7. Symmetry
D(18)	D(64)	E	B	H0	H11	H12 / Others for further study	*Access attributes* 8. Access channel and rate
I.440	I.451	CCITT No. 7	I.462	Others for further study			9.1 Signalling access protocol
G.711	G.721	I.460	I.451	X.25	Others for further study		9.2 Information access protocol
(Note 3)		(Note 4)					
Under study							*General attributes* 10. Supplementary services provided 11. Quality of service 12. Interworking possibilities 13. Operational and commercial

Note 1. The characterization of the information transfer configuration attribute "broadcast" is for further study.

Note 2. The need for a "data sequence integrity" attribute is for further study.

Note 3. The use of Recommendation G.721 as an information access protocol is for further study.

Note 4. The use of Recommendation I.451 as an information access protocol is for further study.

Note 5. A provisional definition of the establishment of communication is given in Recommendation I.130. Further clarification is required.

Note 6. The attributes are intended to be independent of each other.

Figure 4.25 Values for each bearer service attribute (reference 24, Table B-1/I.210).

much of the information. The first edition, indeed, had many more that have been excluded from this edition as they strayed into areas that are more properly dealt with in the relevant recommendations and standards.

A brief introduction to the concept of CCS led to an account of the development of digital CCS in the CCITT signalling system no. 7. The description of #7 relied on other references for the basic transport mechanism and concentrated upon the user parts presently defined for #7. The potential of #7 to transport any message content was insisted upon and contrasted with the actual development that appears to be confining #7 signalling to the business of the administration only. Discussion of the user parts highlighted the areas where the generality of the #7 concept has been eroded by implementation decisions. The treatment of user parts concluded with CCITT DSS 1 (Q.920–Q.931) and led on to brief descriptions of the UK-specific digital CCS user parts DPNSS and DASS 2.

The OSI model was introduced by the narrative annex to the ISO standard. The remaining treatment was fairly brief, relying upon the concrete examples of applied OSI that have already been discussed in dealing with packet switching, ISDN access protocols and #7 signalling. For comparison, a similar interconnection arrangement that was devised before the OSI model, that of TCP/IP, was also described. This has its importance in the treatment of B-ISDN because of its popularity in LANs that are candidates for interconnection via the B-ISDN.

Finally, having avoided using the attribute method of characterizing ISDN services and networks, the chapter ended with a description of the method that is central to an understanding of the information presented in the I-series recommendations on the ISDN.

CHAPTER 5

Broadband ISDN (B-ISDN)

Hardly had the term ISDN been invented and described than the industry added a letter and began talking about the broadband ISDN. It would appear, in retrospect, to be unduly ambitious to conceive of a broadband network when the whole of the 1980s and, it seems, most of the 1990s will have been spent in deploying an inadequate ISDN capability.

The suggestion for broadband first came from the German network provider, at that time the Deutsche Bundespost (DBP), whose approach has consistently been to envisage a broadband ISDN as the ultimate goal and therefore sees little value in offering an intermediate ISDN service to the public over the existing network. To this end, the DBP introduced a broadband trial on the local network, called Bigfon, in 1984. Although ISDN has been generally seen as a means of using the existing local network more effectively, the opposing view, typified by the DBP approach, sees the services possible via ISDN, lacking video, high-speed document transfer, and computer-aided design dialogue, as being quite inadequate. For the Bigfon experiment some 300 subscribers' premises on a new estate in Berlin were equipped with optical fibre distribution providing voice, data, videophone and three simultaneous TV channels to the subscriber. The German strategy was to equip new housing and office developments with optical fibre local distribution as part of the services provision.

Opinions in favour of a rapid transition to a B-ISDN tend to be most strongly held in countries that have deployed cable TV (CATV) systems fairly widely. The cable TV network represents an existing wideband local network that constitutes a viable alternative to the local telecommunications network and could possibly provide B-ISDN access capability. There is a potential conflict here with direct broadcasting by satellite (DBS), which is a direct competitor of cable TV but does not include provision for two-way communications and therefore precludes the provision of interactive services.

The attitude in the UK towards the B-ISDN has changed radically in the past few years as the liberalized telecommunications regime has encouraged the entry into the market of many cable TV operators. The UK therefore is now becoming well placed to offer B-ISDN, at least to those who subscribe to the

112

broadband cable TV networks. The more far-sighted of the CATV operators are indeed installing local networks in fibre using SDH rather than the more conventional coaxial cable despite a significant but decreasing cost advantage for the latter.

Despite this enthusiasm for B-ISDN in some countries and among certain operators, the publication of the necessary international standards has lagged behind even the prolonged delays for ISDN standards. Additional impetus has now been provided, again by Germany, as a result of German reunification. There was an urgent need to replace the quite inadequate telecommunications network in the former East Germany, and this could most easily be achieved by using fibre local distribution throughout. As a result, the European nations set up a special working group to rush the necessary standards[1] describing the V5 access network interfaces into being to enable the rewiring of East Germany to be achieved.

Two methods of providing the large transmission capacities of the B-ISDN in a way that allows flexible allocation of a very wide range of data rates are available. The methods are now seen as being used in co-operation. One is the synchronous digital hierarchy (SDH), which is really applicable only as a transmission system, and the other is ATM, which is applicable to broadband user services. Both of these will be treated in some detail in the next chapters.

The choice between SDH and ATM is really a choice between circuit switching as the preferred method for the B-ISDN and packet switching. In the event, ATM and packet switching have been chosen on the assumption that it provides flexibility, bandwidth efficiency, minimal switch processing and simplicity. In practice there are real problems in congestion control and complex call and path selection procedures that have long been solved in circuit switching but are still in process of solution with ATM. See for example reference 162.

The SDH is a means of loading a 150 Mbit/s "envelope" with a wide variety of different services at a wide variety of data rates and still be able to multiplex and demultiplex individual portions without having to break down the complete envelope. This is done by providing a pointer in a known part of the envelope to indicate the whereabouts within the envelope of each individual communication.

The ATM approach is a development of the packet-switching method using small, equal-size packets, called "cells", that are small and simple enough to be dealt with at the very high speeds involved.

Both methods provide a means of transmitting many, possibly high-bandwidth, communications and accessing any one of them without the need to break down and analyse the complete transmission.

The overwhelming advantage of the ISDN is the provision of a wide range of services via a single network over a single standard access. Thus, the basic access and primary access defined for the ISDN are seen to be fundamental and it is not sensible to add a further set of access interfaces for the B-ISDN.

1. ETSI subtechnical committee SPS 3.

Clearly, new accesses are required to interface with at least the 155.52 Mbit/s SDH or equivalent ATM, and this is the only new access proposed. It is defined for 155.52 Mbit/s and 622.08 Mbit/s. It is defined in terms of ATM but with provision for optional transmission frame adaptation for either cell-based interfaces (ATM) or SDH interfaces. The B-ISDN will arrive at the (very large[2]) user and at the local public network distribution point as a 155 Mbit/s transmission. Within the envelope, services requiring the high capacity that the B-ISDN provides will probably be in ATM format. "Normal" services, including plain old ISDN, will be extracted out of the transmission as 2 Mbit/s systems for presentation at an ISDN primary access or at digital to analogue conversion equipment for onward analogue delivery or as 144 kbit/s systems for delivery to an ISDN basic access. Figure 5.9 summarizes these accesses.

Table 5.1 ISDN and B-ISDN protocol specifications[22, 23, 33–37, 70–77, 88–94].

	ISDN and B-ISDN		B-ISDN	
Layer 1	I.430[22]	Basic layer 1 specification	I.432[85]	Physical layer specification
	I.431[23]	Primary rate layer 1 specification		Including transmission convergence sublayer
ATM Layer			I.361[88]	ATM layer specification
			I.362[89]	ATM adaptation layer (AAL) functional description
			I.363[90]	ATM adaptation layer (AAL) specification
			Q.2100[91]	SAAL, Dec 94
			Q.2110[92]	AAL SSCOP, Dec 94
			Q.2130[93]	AAL SSCF, Dec 94
Layer 2	Q.920[33]	Data link layer, general aspects		
	Q.921[34]	Data link layer, specification		
Layer 3	Q.930[35]	Layer 3, general aspects		
	Q.931[36]	Layer 3, specification for basic call control	Q.2931	B-ISDN equivalent of Q.931
	Q.932[37]	Generic procedures for control of ISDN supplementary services		
	Q.933[70]	Layer 3, specification for frame-mode bearer services		
	Q.950[71]	ISDN supplementary services		
	Q.951 to Q.957 [72–77]	Stage 3 descriptions for ISDN supplementary services		

2. "Very large" is defined in terms of bandwidth use. CATV users with 30 2–6 Mbit/s broadcast channels or a LAN user with a peak data requirement of 100 Mbit/s are not really very large but will require 155 Mbit/s interconnection.

Recent additions to ITU-T Recommendation I.432[85], which describes this interface at the physical layer, are Annexes D and E providing for interfaces at 1.544 and 2.048 Mbit/s. Thus, the simple initial perception of just one additional interface for the B-ISDN has had to be extended to make the B-ISDN available to the "small user", in contrast to our definition of the very large user in the previous paragraph.

The approach used by ITU-T in defining the B-ISDN has been to include B-ISDN aspects in the existing recommendations where this is appropriate and to prepare additional recommendations only where this is absolutely necessary for topics that are peculiar to the B-ISDN alone. Table 5.1 is an extended version of Table 4.1 showing the additional B-ISDN protocol recommendations and their relationship to the existing recommendations for the ISDN.

5.1 Why broadband?

The introduction above has attempted to answer this question, at least from the point of view of network operators who already have broadband capability. As part of our introduction to the broadband aspects of ISDN it may be helpful to approach the question from several different directions.

5.1.1 Wideband and broadband

The first edition of this book showed the difficulty involved in the, supposedly rare, occasions when a channel capacity of more than 64 kbit/s was required for an ISDN connection. Some of that argument is reproduced here as an introduction to the problem of providing broadband in existing networks.

Wideband is used here to describe services made available over the ISDN access as presently defined that require a greater bandwidth than the access bandwidth, i.e. greater than 64 kbit/s. Wideband could include such things as high-speed, high-capacity, data transfer or video telephony, possibly of a reduced picture quality or at slow speed. To provide wideband over the basic and primary access means combining, for the duration of a call, several B channels of a primary access or the B1, B2, and perhaps the D channel of the basic access. Thus, although a normal ISDN call is made to the network address of a single channel, a wideband call must be made to several addresses at once and the resulting total channel capacity must be used in aggregate. The problem is partly one of switching. Each ISDN access channel is normally treated in isolation – separate network address, separate connection. To combine several network addresses in the one call and use their total channel capacity requires the switching instructions used to take account of this. The problem is illustrated in Figure 5.1.

Figure 5.1 shows an exchange that makes no particular effort to switch related channels in any particular way. Assuming that the wideband connection is to be between network addresses incoming system 1, channels 8 and 9, and outgoing system 2, channels 27 and 28, then the difference in arrival times of signals launched from 1/8 and 1/9 at the same moment is 125 µs. In this example, using "classical" TST switching, the possible delay between the related channels is limited to zero or 125 µs. The additional limitation involved in requiring related channels to be switched in a constrained way is, therefore, that both must be given an intermediate channel; this scan or both must wait for the next scan. The blocking probability experienced by a 128 kbit/s call with this constraint will be considerably more than twice the blocking probability of a normal 64 kbit/s single channel call.

The relationship assumed in Figure 5.1 between the network addresses of the related channels that might require wideband switching may also be difficult to realize in practice. The fact that the classical TST network was chosen for the example in Figure 5.1 identified a relatively simple, although serious, constraint on wideband calls. Had a more sophisticated switching system example been chosen, the constraint would be more onerous.

There is no doubt that wideband switching can be implemented in a number of ways. What has been said, however, is enough to indicate that its existence in the ISDN, in any but negligible proportions and despite the essentially non-blocking nature of digital switches, could have an overriding effect on the provision of the network switching components.

Transmission problems are also associated with wideband capability. It is a waste of time maintaining time slot sequence integrity across the switch if the circuits used have different propagation times. It may well be that a simpler solution to the wideband problem would be to ignore its existence within the

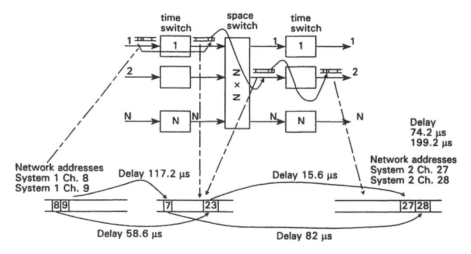

Figure 5.1 Wideband switching – demonstration of failure.

116

Table 5.2 Plesiochronous digital hierarchies.

		North America		Rest of the world		Japan	
	Code	Bit rate (kbit/s)	Voice channels	Bit rate (kbit/s)	Voice channels	Bit rate (kbit/s)	Voice channels
Sub-64		2.4		2.4			
kbit/s		4.8		9.6			
		9.6		19.2			
		56.0					
Level 0	DS0	64	1	64	1	64	1
Level 1	DS1	1544	24	2048	30	1544	24
Level 1+	DS1C	3152	48				
Level 2	DS2	6312	96	8448	120	6312	96
Level 3	DS3	44736	672	34368	480	32064	480
Level 4	DS4E	139264	2016	139264	1920	97728	1440
	DS4	274 176	4 032				
Level 5				565148	7680	397200	5760

network and solve the problem at the network terminations. The network termination would include a wideband switching box that would provide aggregate rate, consistent input and output from the variable rate contributing bearer channels.

5.1.2 Synchronous and plesiochronous transmission[3]

Recent developments influencing, in the first place, digital leased circuit networks have been the provision of high capacity digital transmission media using fibre optic technology. This has necessitated interest in cross-connect, not only at the 64 kbit/s and subrate levels, but also at the system levels of the higher order multiplexes. The multiplex hierarchies in use are listed in Table 5.2.

As primary PCM systems (2 Mbit/s, or 1.5 Mbit/s in North America) began to saturate the network, transmission hierarchies were developed to multiplex traffic to higher rates (Table 5.2). Each primary system has its own independent clock, which results in slight differences in frequency. This is known as plesiochronous transmission. To compensate for this, padding bits are added at each multiplexing stage. With the introduction of digital switching however, primary rate systems have become largely synchronous to each other, but the higher order bearers are plesiochronous and independent.

Although these arrangements were well suited to the transport of bits there remained a major drawback. The confusion of multiplexing different levels with their padding bits made it impossible to identify and extract an individual channel within a high-capacity bit rate link. The complete demultiplex procedure had to be carried out to gain access to a particular channel.

---

3. A good account of the background to the development of the SDH is contained in reference 95.

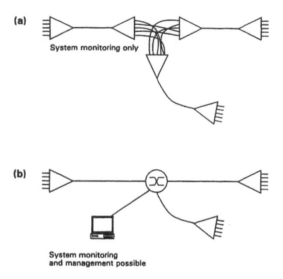

Figure 5.2 Development into synchronous networks.
(a) Current situation. (b) SDH or SONET.

This situation was acceptable when only point-to-point links that were fully terminated at either end were involved. Applications such as drop and insert were not viable. As networks developed and became more complex, the interlinking of traffic at nodes required banks of multiplexers and large distribution frames or digital cross-connects (DCCs). Figure 5.2a illustrates the situation. It became clear that the existing multiplexing standards, designed for point-to-point links, were no longer suitable for large meshed networks.

As a result, a number of bodies, including AT&T, Bellcore and BT, developed alternative approaches in which channels were arranged synchronously within the higher order multiplex. Low-rate channels could then be extracted without going through the demultiplexing process. Figure 5.2b illustrates what this might do to the network.

5.1.2.1 Synchronous digital hierarchy (SDH)
New proposals, based on this perception, were driven initially by North American requirements and were expressed in the North American standards for synchronous optical network (SONET). In February 1988, however, agreement was reached at CCITT for a world standard. This standard was published in the Blue Book as Recommendations G.707, G.708 and G.709 under the heading synchronous digital hierarchy (SDH)[96, 97, 163].

The North American network is using SONET, and it will be necessary to remember the major difference between SONET and SDH. The SONET envelope permits a bandwidth of 51.84 Mbit/s although the SDH bandwidth at STM-1 is 155.52 Mbit/s. The two structures are shown in Table 5.3.

Table 5.3 SDH and SONET rate structure.

SDH		SONET	
Synchronous transport module	Bandwidth	Level	Bandwidth
		OC-1	51.84 Mbit/s
STM-1	155.52 Mbit/s	OC-3	155.52 Mbit/s
STM-2	311.04 Mbit/s		
		OC-9	466.56 Mbit/s
STM-4	622.08 Mbit/s	OC-12	622.08 Mbit/s
		OC-18	933.12 Mbit/s
STM-8	1.24416 Gbit/s	OC-24	1.24416 Gbit/s
STM-16	2.48832 Gbit/s	OC-48	2.48832 Gbit/s
STM-64	≈10 Gbit/s proprietary	OC-192	≈10 Gbit/s proprietary

The SDH (or SONET) standard has five important features:

(a) It is synchronous, which allows efficient drop and insert and cross-connect applications.

(b) It is an optical standard ensuring compatibility at the optic signal level between equipment from different manufacturers.

(c) It makes provision for network management channels ensuring effective control and reconfiguration of networks.

(d) It can be introduced directly into existing networks in both the 24-channel and 30-channel "worlds".

(e) It is being formulated at the same time as the B-ISDN requirements and can embrace these within the SDH standards.

The philosophy underlying the standard is that any of the currently used transmission rates can be packaged into a standard-sized container and located in an easily identifiable position in the multiplex structure.

For all the transmission rates in use there are (at least proposed) mappings into containers, called virtual containers (VCs). Once located in containers, multiples of different containers can be combined together into a standard format. In this way the structure can be used for 2 Mbit/s, 8 Mbit/s, 34 Mbit/s and 140 Mbit/s traffic or for North American 1.5 Mbit/s, 6 Mbit/s and 45 Mbit/s traffic. Containers can be mixed, allowing 1.5 Mbit/s and 2 Mbit/s traffic to be carried simultaneously within the same structure.

To achieve efficient mapping within a 140 Mbit/s structure and to provide sufficient management overhead, 155.52 Mbit/s has been chosen as the basic SDH rate. This is known as synchronous transport module 1 (STM-1) and forms the building block for higher rate traffic. Higher rates can be formed by simple multiplexing (Table 5.3).

5.2 Towards faster data transfer

Digital transmission and switching has resulted in a convergence of the skills of the telecommunications switching engineer and the telecommunications transmission engineer. However, at the same time that the transmission experts were solving the problems of accessing a small part of a broadband transmission, the switching experts were solving the analogous problem of transmitting and delivering large bandwidth signals to the subscriber's terminal. Thus it is that SDH makes possible access to any part of a high-bandwidth transmission but does nothing to simplify delivery to the subscriber; ATM permits high-bandwidth sending and delivery but is not necessarily ideal for transmission across the network. In other words, technologies able to solve the problem of high-bandwidth telecommunications have been developed, but the solutions are not necessarily what would have been proposed had the whole problem been specified for the designers at inception of the design.

Packet switching was described in Chapter 2. How it influenced the development of ISDN and how, in the ISDN, much data is transferred using protocols based upon the X.25 packet-switching protocol was discussed in Chapter 3. The limitations of X.25 where it is required to send information, such as voice, that would be destroyed by very long or frequent delays have also been mentioned. The search for a better way of transmitting information of this kind from user to user has led to the development of methods of fast packet switching. Among these, frame-mode bearer services will be discussed briefly before considering the eventual candidate for the B-ISDN: ATM.

5.2.1 Frame-mode bearer services[98, 99]

Packet switching using X.25 is an example of an "enveloping" information transfer protocol (see Ch. 2). The data to be transmitted is enclosed in successive envelopes of signalling information directed at the various layers in the information transfer hierarchy, the OSI model. One disadvantage of the simpler forms of these protocols, such as packet switching, is that the signalling and the data are transmitted as one and subjected to the same error correction and retransmission mechanisms and the same flow control mechanisms so that the data transfer is no longer transparent from user to user.

As the amounts and the speed of the data increase, there is a more urgent need to separate the data from the signalling. This enables more bursty kinds of data, such as variable bit rate video, and much higher bit rates to be transmitted than is possible in enveloping protocols.

To meet these requirements, ISDN is designed to separate, to an extent, the signalling on the C-plane from the user information on the U-plane (see Fig. 2.12). ISDN frame-mode bearer services follow this principle. Virtual calls are set up and controlled using the same outband ISDN signalling procedures, unlike X.25, where signalling and data are carried inseparably in the same logical channel.

Two distinct ISDN frame-mode bearer services have been defined: frame relaying and frame switching. The same signalling procedures are used for both but they differ in the protocol supported by the network in the U-plane during information transfer. In either service the user is able to set up a number of virtual circuits simultaneously to different destinations.

5.2.1.1 Frame relay[100]

The simplest frame-mode service, frame relay, is shown in Figure 5.3, which should be contrasted with Figure 2.12. Signalling in the C-plane uses the normal ISDN signalling for layers 2 and 3 with some enhancement to support the frame-mode parameters. In the user plane, however, the network supports only a part of the link layer (layer 2) protocol referred to as the "core" functions, which are: separating the frames using HDLC flags; forcing data transparency by zero bit insertion and extraction; checking frames for transmission errors and length (any frames found to be in error are discarded) and multiplexing and demultiplexing the frames associated with different virtual calls on the basis of their layer 2 address. Multiple virtual calls to different destinations can be supported simultaneously.

The basic service provided by frame relay is the unacknowledged transfer of frames between ISDN terminals.

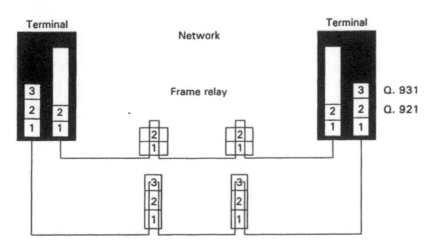

Figure 5.3 Frame relay.

5.2.1.2 Frame switching

The protocol architecture for frame switching is shown, in the same format, in Figure 5.4. Again, signalling in the C-plane follows ISDN practice, but in this case the network operates the complete link layer protocol in the U-plane

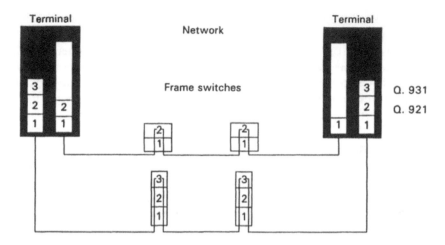

Figure 5.4 Frame switching.

during information transfer. Frame switching therefore provides acknowledged transfer of frames between terminals. The network detects and recovers from lost or duplicated frames and frames containing errors and it will operate network-enforced flow control. As with frame relay, multiplexing, demultiplexing and routing at each switching node are done on the basis of the layer 2 address. Again, higher layer protocols are operated by the terminals on an end-to-end basis.

The use of frame-mode services considerably enhances the ability of the ISDN to carry a wider variety of applications. Allied to the broadband ISDN it may provide a vehicle for the transmission of voice and video services without the need for circuit-mode connections. The lightweight frame relay protocol supports applications (such as voice) where additional delays cannot be tolerated. Frame switching offers the robustness against errors usually associated with X.25 but without the per-packet overhead of layer 3.

5.2.1.3 Packet-switching and frame-mode services[101, 102]

Table 5.4 summarizes the functions supported by packet switching and frame-mode bearer services. X.25 uses, at layer 2, as we have seen, the balanced link access procedure (LAPB) of the HDLC protocol. In LAPB the links between nodes and terminals and the internodal links perform all the layer 2 functions as defined for the OSI model. These functions are frame boundary recognition using the flag and the related bit stuffing to guarantee bit transparency. In addition, it performs CRC generation at the sending end and CRC check at the receiving end together with retransmission of missing or errored frames using an automatic repeat request (ARQ) protocol. X.25 also performs flow control and multiplexing of logical channels at layer 3.

In frame relay, retransmissions of user data frames for error correction

122

Table 5.4 Functional differences between packet-switching systems.

Function	X.25	Frame switching	Frame relay
Frame boundary recognition (flags)	X	X	X
Bit transparency (bit stuffing)	X	X	X
CRC generation and check	X	X	X
Error control (ARQ)	X	X	
Flow control	X	X	
Multiplexing of logical channels	X		
Maximum practical data rate	2 Mbit/s	8 Mbit/s	140 Mbit/s

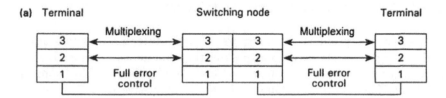

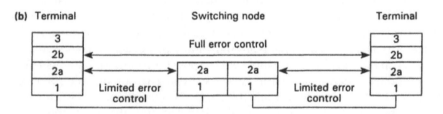

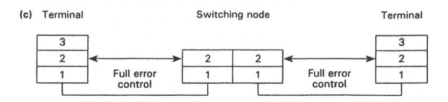

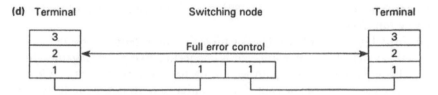

Figure 5.5 Packet switching and related protocols. (a) Full error control on every link, packet-switched network. (b) Limited error control in frame relay networks. (c) Full error control in frame-switching networks. (d) Cell switching in ATM networks.

purposes are only performed end to end, between the user terminals, and not link by link. Between switching nodes an error detection function is performed to only discard errored frames that it is a waste of time to pass along the network. There is no flow control and no multiplexing at the packet level.

In frame switching, error control and flow control are retained and continue to be performed on a link-by-link basis.

Table 5.4 shows the approximate maximum data rate achievable using these protocols at the present time. Figure 5.5 illustrates the differences in these methods and ATM in terms of the OSI model.

5.2.2 Asynchronous transfer mode (ATM)

SDH solves the problems of transporting many different communications within a high-capacity network by packing the communications together and using a fairly sophisticated pointer mechanism to identify individual communications. The pointer mechanisms, however, only have to operate at low speeds of a few kbit/s and not at the interface rate.

ATM, which began as an alternative method of performing the same task, has come to be regarded as a co-operating technology which, in conjunction with the SDH, delivers high-capacity communications over the B-ISDN. At the conventional data rates of the ISDN and with existing facilities, communication in real time, say by voice, would only be possible by using circuit switching to guarantee a permanent path. Packet switching is not suitable for voice because the delays involved would destroy the conversation. ATM is a form of packet switching that uses small, equally sized packets, called cells, and a protocol that is sufficiently simple to permit cells to be transmitted, interpreted and delivered fast enough for them to carry any kind of information including voice and video. The structure of the ATM cell is shown in Figure 5.6.

Using ATM, a wide range of channel rates can be accommodated on a single bearer. The control mechanism must, however, operate at the bearer rate. ATM uses short, fixed-length cells with minimal headers to allow calls to be routed at high speed by means of hardware-implemented routing tables at each switching node. International agreement is that the header of each cell will consist of five 8-bit bytes and that the cell information field will consist of 48 bytes, making a total length of 53 bytes. The header provides two main routing functions:

(a) Virtual path identifier (VPI). A path is the equivalent of a route in a circuit-switched system, permanently connecting two points together. In an ATM environment, the path would not have a fixed capacity. "Virtual" means that cells can be routed from source to destination on the basis of the VPI in whatever way seems appropriate.

(b) Virtual circuit identifier (VCI). Calls set up as required over the virtual path indicated by the VPI.

124

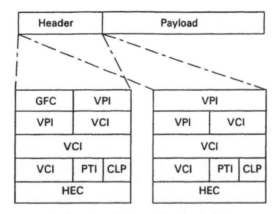

Cell header at UNI

Cell header at NNI

An ATM cell has a fixed length of 53 bytes, or octets, divided into a 5-octet header and a 48-octet information field (payload).

The main function of the ATM cell header is to route the cell from the point of origin to the point of destination. The routing information is contained in the VPI and VCI fields. The header is slightly different in the UNI compared with the NNI. The UNI contains 4 bits of GFC (generic flow control). It will be used to ensure fair and efficient use of available capacity between the terminal and the network.

The payload type field containing the payload type identifier (PTI) is used to indicate whether the cell contains user information or special network information, for maintenance, for example.

The cell loss priority field (CLP) indicates a two-level priority and is used if it becomes necessary to discard cells, depending on network conditions.

The header information is protected by a check sum, which is contained in the HEC (header error control) field.

Figure 5.6 ATM cell structure.

The separate development of ATM and SDH is emphasized by the use of the same abbreviation for different concepts. In SDH the VC is the virtual container into which different kinds of transmission are packaged. In ATM, as can be seen, the VC is the virtual circuit, part of the addressing mechanism.

ATM can flexibly handle all types of traffic. Examples are:

(a) *Voice*. 64 kbit/s voice can be assembled into ATM cells. Each cell, with 48 bytes of information in it, can contain 6 ms of speech. For some purposes the delay this introduces, called the packetization delay, may be excessive, in which case only partially filled cells or multiplexing within cells may be used.

(b) *Video*. Using modern techniques of condensed video, only changes in the picture signal are transmitted. This information, detailing the variation with the activity in the picture, results in a video signal to be transmitted that is extremely variable in content and is ideally suited to the variable capacity features of ATM.

(c) *Signalling*. Signalling is based on the existing ISDN protocols but assembled into cells. Some of these cells would have to be given high priority, necessitating the use of flow control techniques in ATM.

125

5.3 The access network

It has already been emphasized that the ISDN is predicated upon integrated access, a concept that is perhaps much more central to the technology than the integrated network itself. This was a realization that became apparent in the industry as the concept of the ISDN was developed – it was not present in the original idea. In a similar manner, the concept of the access network has become important, perhaps central, to the B-ISDN as the broadband concept has developed although it was not obviously present at the inception of the idea. Thus, the first edition of this book made much of the concept of integrated access, but in the little that was said about broadband there was no mention of the access network as a specific issue because neither the author nor the industry was aware of it as an issue.

While discussing broadband then it would be useful to revise a little of the history that has brought the concept of the access network into prominence.

Figure 3.1 illustrated how development has caused the intelligence of the network to migrate out from the switching nodes towards, and eventually into, the subscriber's terminal equipment. Like all attempts at summary it is not the whole story. The increasing intelligence of the terminal has demanded increasing intelligence from the network, or at least from the services provided via the network. Thus, the concept of the intelligent network (IN) has been introduced, defining a much higher degree of intelligence resident, no longer within the network, at central service nodes available to users via the network.

As it appeared in the first edition, Figure 3.1 did not emphasize the fact that along with the movement of intelligence towards the subscriber there was an accompanying move of switching functions out of the telephone exchange into the local access network. Figure 3.1 is reproduced here in a modified form as Figure 5.7 to emphasize how the concentration stage of the exchange is now located, more often than not, remote from the exchange in a roadside cabinet or, perhaps, in the basement of an office block. Indeed, with the provision of Centrex service, the concentration stage may be not only remote from the exchange but also incorporate the remaining localized functions of the private exchange.

We mentioned earlier the development of the V5 standards[103, 104] to permit the introduction of modern methods into the local network, initially driven by the need to provide a rapid improvement to the local networks in East Germany and other former Communist countries. One reason for the urgent need for these new standards is illustrated in Figure 5.7. As the intelligence of the network migrates towards the subscriber, represented in the figure by the remote concentration stage, more of the local network is provided of necessity by the same supplier as the supplier of the local exchange. This is because all the switching systems incorporating the facility for remote concentration stages use a proprietary interface between the local exchange and the concentrator.

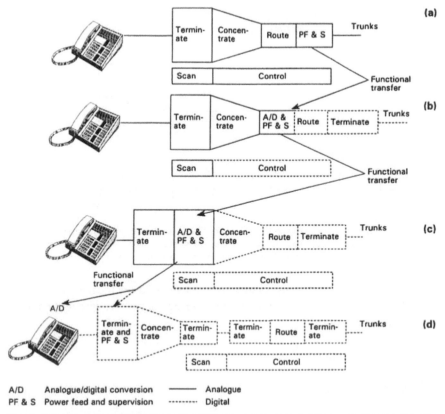

Figure 5.7 Transition stages towards the ISDN: development of the access network. (a) Analgoue environment. (b) Digital environment for concentrated traffic. (c) Digital environment within switch node. (d) Totally digital environment with remote concentrators – the ISDN.

In setting the objectives for the V5 standards it was thought important to abolish this technical opportunity for monopoly and define, not only a user–network interface (UNI), as we have already done for user access to the ISDN, but also the service node interface (SNI). Figure 5.8 illustrates the progression towards these interfaces.

This more careful definition has introduced a new nomenclature into the access network, besides renaming what we had understood as the local network or perhaps the local access network. The new terms include:

(a) SNI (service node interface). In discussing the ISDN this has been referred to as the V interface (hence the term V5 standards); it is also sometimes (and erroneously) called the network node interface (NNI).

(b) UNI (user–network interface). This term is already in use in the ISDN recommendations although in this book the term the U interface is also used to mean that part of the UNI that physically connects to the wires coming into the user's premises.

127

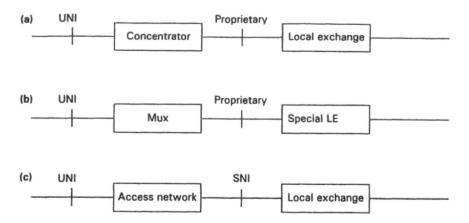

Figure 5.8 Methods of realizing the access network. (a) Switching system using remote concentration stage. (b) Switching/transmission system with remote equipment, e.g. multiplexers. (c) Access network as defined in the V5 standards.

(c) SN (service node). This is what used to be called the local exchange (LE).

(d) NNI (network node interface). The interface between switching nodes or between networks.

Chapter summary

The broadband method

We have seen how the desire for broadband communications to extend the usefulness of the ISDN came about. We have seen how two different methods have been developed to address the problem of transmitting high-bandwidth communications while being able to access any particular part of that communication and we have seen how SDH performs this in an optimum fashion for the transmission engineer but does little to help the user's terminal to place high- and variable-bandwidth communications on to the network and to receive communications from the network. ATM, on the other hand, permits the user to have high- and variable-bandwidth access, but the high-capacity ATM stream across the network is not necessarily best suited to carry other forms of transmission, particularly constant bit rate transmissions such as 2 Mbit/s or higher order plesiochronous PCM streams. And we have seen the increasing importance attached to defining the access network in such a way that there is a free choice in selecting the component parts of the access network.

Although frame-mode bearer services were discussed, and although they are defined by ITU-T as candidates for the ISDN, it was shown that their development and more widespread use have probably been overtaken by events in the form of ATM. Little more will therefore be said about frame mode. The account has served to put the historical development of SDH and ATM into context.

Thus, the pattern of development has been to envisage a mixture of the two techniques with the existing techniques of the ISDN. We have identified that the B-ISDN will access the user using one of a selection of interfaces, some of which may not yet have been defined, and these, at least those that we know about, are shown in Figure 5.9.

Evolution to the B-ISDN

The similarities in the plot between the story told in the first edition and the story of the introduction of the B-ISDN are remarkable. The early chapters of this book explained how the ISDN was conceived because of a perception that the digital network permitted the introduction of a combination of digital transmission and switching, a solution that offered tremendous cost savings by reducing the variety of networks required to just one: the integrated network. To achieve these savings it was necessary to persuade users to provide for themselves the digital terminal device and, to help to persuade them, the increased functionality of the integrated services network was offered as an incentive.

In fact, things did not work out that way. No network provider offered the necessary guarantees that would give manufacturers the confidence to

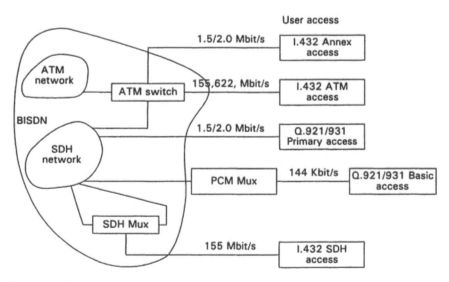

Figure 5.9 Broadband ISDN access methods.

develop the new terminal devices. Indeed, in developing the ISDN standards the network providers insisted on including means to collect payment for each individual integrated service, thus removing much of the attractiveness of the ISDN to users. And alongside the ISDN development much of the new functionality that it would provide was made available anyway through other networks or other, often more expensive, means.

In parallel with this tale of great hopes dashed by small imagination, and confusing the putative ISDN developer still further, was the developing, and much publicized, argument that the ISDN was insufficient and a broadband version must supersede it.

Figure 5.9 does not exhaust the interface possibilities. ITU-T Recommendation G.804[105] is intended to provide transport for ATM cells during the transition period while SDH is being deployed. The recommendation provides the mapping to be used for ATM cell transport of the different PDH bit rates for both 1.544 Mbit/s and 2.048 Mbit/s hierarchies. Because of some alarm in the industry concerning the cost of SDH, this transition period may be prolonged.

CHAPTER 6

B-ISDN tools: ATM

6.1 Introduction

Two different technologies were introduced in the previous chapter, conceived and developed to meet widely differing needs, both of which turn out to be suitable for use in the implementation of the B-ISDN. Today, with hindsight, many would argue that one of them, SDH, although being very relevant to the solution of problems of transmission, is not part of the B-ISDN toolbox. Communications initiated by B-ISDN users may well be transported using SDH, but this is irrelevant to the development of the B-ISDN technology.

The author's view is that this very fact, that much of B-ISDN communication, along with much "ordinary", conventional communication, will take place using SDH transport, qualifies SDH as a required topic in any discussion of B-ISDN. Furthermore, B-ISDN is not finished – it is too early to say that we have the answer to all our broadband communications problems, and SDH and technologies using SDH ideas may well still have an important, direct part to play.

This chapter therefore will be devoted to ATM, describing the underlying concepts in rather more detail than was attempted in the brief introduction of the previous chapter. SDH will be treated similarly in Chapter 7.

To give a little background to the discussion, Table 6.1 indicates a brief history of the developments that have led up to the implementation of the B-ISDN and, specifically, of ATM.

6.2 Asynchronous transfer mode (ATM)

6.2.1 The ATM development story[1]

Practically all that has been said about the telecommunications network in the introductory sections of Chapter 2 and in describing the ISDN in Chapters 3

---

1. Much of this account is based upon an excellent but rather inaccurate article [106].

Table 6.1 Development milestones for B-ISDN and ATM.

Mid-1960s	AT&T begins developing PCM TDM transmission systems for the public network
1967	First introduction of T1 PCM transmission systems
Late 1960s	Bell Laboratories investigates blending label-based switching and packet switching with TDM
1967	Bell Laboratories introduce the term asynchronous time-division multiplexing (ATDM) to describe a cell relay proposal
1971	Work begins at CCITT on technologies that could form the basis for the ISDN
1972	Ethernet introduced by Xerox Corporation, IBM introduces synchronous data link control (SDLC).
1974	ISO unveils the OSI model
1974	IBM responds with systems network architecture (SNA)
1976	CCITT adopts the X.25 protocol for packet-mode communications
1980	A CCITT recommendation is issued describing the ISDN
1984	CCITT recommendations are published covering the ISDN in a reasonable degree of detail
1986	The CCITT decides to make ATM the core technology for the B-ISDN
1987	ANSI adopts the first standards describing the synchronous optical network (SONET)
1988	CCITT recommendations are published covering the B-ISDN. Included are Recommendations G.707, G.708 and G.709 on SDH
1991	ITU-T issues first recommendations specifying ATM for B-ISDN
1991	Foundation of the ATM forum
1991	Stratacom offers the first commercial (non-standard) ATM product; the IPX switch

and 4 has been based upon the current reality of a network that is founded upon circuit switching. This is despite the fact that the message-switching concept of the X.25 packet-switching protocol has been used as the basis of the ISDN user–network interface. The digital network of the present day upon which the ISDN is based is a synchronous, circuit-based network. As the quantity of data carried by such a network increases, the inherent inefficiency of allocating 64 kbit/s to every communication regardless of its actual requirement in bandwidth, which for a great deal of data may be much less than this, becomes an important consideration.

In considering the needs of the B-ISDN we are both extending the amount of data and data-related traffic that is to be carried and increasing the maximum bandwidth that it is possible to carry. Thus, in a circuit-switched environment, the problem of the variable bandwidth requirement introduced in the previous chapter under the heading of wideband switching is greatly magnified.

The definitions of what is to be required of the B-ISDN assumed that an important element would be video, and the received wisdom of the time saw this as a requirement to carry a broadcast-quality video signal in about 140 Mbit/s. Subsequent advances in compression techniques have probably made this too liberal a target, but it would be unwise perhaps to reduce the requirement as in all other respects the world seems to require ever-increasing allocations of bandwidth.

Development therefore turned to the concept of packet switching. Rather than allocating a circuit consisting of a defined constant bandwidth for the duration of the call, divide the information into packets, label each packet with source and destination and switch and transmit the individual packets across the network in whatever seems to be the most acceptable way. This reduces the bandwidth inefficiencies but introduces the problems of variable delay distortion to the reconstituted message. For relatively continuous and even messages such as voice and video signals these distortions, even using very fast packet switching, would be intolerable unless steps are taken to limit, perhaps standardize, the size of the individual packet. For data, as well as for voice and video, the packets must, at least, arrive in the same order in which they were sent.

The previous chapter described attempts to increase bandwidth by using frame relay and frame switching. That discussion outlined why it was thought better to revert to an older idea, devised by W. W. Chu in Bell Laboratories and named asynchronous time-division multiplexing (ATDM), of splitting the message up into cells of equal length, each with a standardized header containing brief indications of the nature and destination of the message. Thus, the basic form of the B-ISDN defined by CCITT in 1988 consisted of a combination of ATDM, renamed ATM, and SDH, providing a cell-based user–network interface with transmission over a synchronous network based upon SDH at 155 Mbit/s. SDH was itself a redefinition of the synchronous optical network (SONET) already defined in the US and using a basic transmission speed of 51 Mbit/s.

An immediate problem in defining ATM was the choice of the size of the ATM cell. The data networking industry, initially averse to the idea of chopping up data into fixed lengths, began work on ideas for fast packet switching and eventually plumped for a cell length of 128 8-bit bytes. The telephone industry, on the other hand, as a result of work at Bell Laboratories and CNET, saw difficulties in packing voice traffic into anything larger than a cell of 16 bytes.

The CCITT, renamed ITU-T, recommends that the maximum tolerance for delay on terrestrial voice circuits is 20 ms. Beyond this limit echo suppression or echo cancellation should be used on the circuit. A voice conversation coded at 64 kbit/s will fill a 16-byte cell in 2 ms. Filling a 128-byte cell, however, will take 16 ms and, were more efficient coding schemes in use, the delay would be even longer. Thus, the telephony network operators foresaw a much greater need for echo cancellation. This might not be a problem in North America, where the distances to be covered demand echo cancellation in any case, but would represent a new source of additional expense in Europe, where circuit lengths are short enough to preclude the need for measures against echo in most cases.

After a great deal of discussion the telephone companies offered to raise the limit to 32 bytes and the data industry offered 64 bytes. At this point the CCITT stepped in and split the difference. The cell length would be 48 bytes plus a five-byte header. No one was happy.

Around this time, the early 1990s, the ATM idea was seen to offer a new alternative to the problems of increasing the capacity of LANs. Ethernet, by then, had become by far the most popular LAN protocol but was proving increasingly inadequate to the demands for capacity on the LAN. Although its capacity is notionally 10 Mbit/s, once control and management are taken into account an Ethernet LAN is unlikely to provide a real capacity for data in excess of 4 Mbit/s. The emerging alternative was to provide a central fibre-based ring using FDDI working at 100 Mbit/s or, more recently, 200 Mbit/s, feeding a number of subsidiary Ethernet or other LAN protocol networks. Such an arrangement is described in Chapter 2.

The combination of FDDI and Ethernet, although being a neat solution to the problem, does introduce the additional complication of an increased number of interfaces. The possibility of a high-speed protocol capable of working from user to user across the whole of the high-capacity LAN caused ATM to be considered, found attractive and implemented as a LAN protocol. ATM will, of course, have to compete with upgraded Ethernet, several versions of which are already capable of offering 100 Mbit/s capacity, and Ethernet and FDDI combinations. ATM does, however, have the advantage that it can carry traffic seamlessly user to user over combinations of LAN, WAN and B-ISDN public network.

6.2.2 ATM: the concept

In circuit switching, a connection is established for the duration of the call between the two end users, providing what is normally a transparent path from user to user. Such a service is an example of a connection-oriented service. Although information may not be passing all the time on such a service, the connection remains established and the customer pays for the service throughout the duration of the call. The structure and the content of the information passing is of little or no interest to the carrier.

By contrast, in the X.25[15] packet protocols, the information to be carried is assembled into discrete packets, each packet is prefixed with header information indicating where it is to go, and the communication consists of a number of these packets (which may well vary in length), each of which could, in theory, be sent over different routes. Such a service is an example of a connectionless service as no permanent circuit is established for the duration of the call. It has already been seen that actual practical examples of such a service impose some constraints on the routing of the packets to ensure that they arrive at their destination in the order in which they were sent and that the variable delays possible with such arrangements are kept within acceptable limits. In X.25 however, the routing information is sent at layer 3, so that the layer 2 protocols have to be terminated at each stage so that layer 3 can be interrogated.

134

ATM is a compromise between these two techniques. The information to be sent is assembled into a number of relatively short, fixed-length, packets, called cells (see Fig. 6.1). At the start of the call the information in the header of the leading cell is used to set up a virtual connection and all subsequent cells identified as belonging to the same call are routed over the same virtual circuit. Thus, the routing and switching nodes are expected to maintain details of the ATM call in their memories for the duration of the call. This is a compromise between the underutilization of the circuit-switched connection when carrying sporadic information and the time-consuming processing of the X.25 address information. The disadvantage is that each node is required to remember information about the call during the duration of the call and a fixed-length packetizing delay is introduced into every communication regardless of the total amount of traffic in the network.

As discussed in Chapter 5, the plan for the B-ISDN is to use the existing ISDN user–network interfaces as far as this is possible. Nevertheless, in describing the ATM concept above, it is clear that, having defined a cell structure, it is necessary to define protocol mechanisms for inserting the communication into the cells together with the information necessary to transmit it across the network. To do this requires rather more than the protocols specified for the DSS 1 interfaces. It is necessary then to interrupt this account of ATM with a more detailed description of the strategy defined for aligning the broadband techniques, of ATM and SDH, with the existing user–network interface definitions.

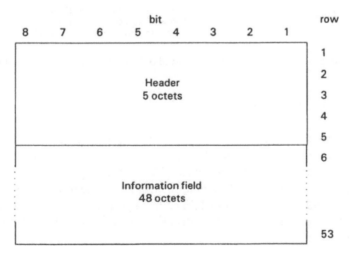

Figure 6.1 ATM cell structure.

6.3 The B-ISDN user–network interface

The equivalent of Figure 3.6 for the B-ISDN is shown in Figure 6.1[107].[2] ISDN presents interfaces to the customer able to support services up to 64 kbit/s. In the same way, B-ISDN must present a minimum set of well-defined interfaces that support all services up to 150 Mbit/s. In Europe, the T_B reference point (and the T reference point in ISDN) also provides the regulatory boundary between liberalized customer premises equipment and the public network. Considerable effort has therefore been expended on the definition of this interface.

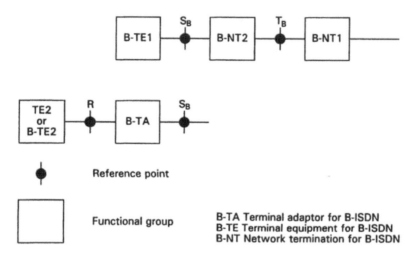

Figure 6.2 Customer access to services supported by a B-ISDN (Figure 1 of Recommendation I.413).

6.3.1 The protocol reference model

To help in the specification of the interfaces at the T_B and S_B reference points, the ITU-T, among others[3], has developed a generic, layered architecture to assist in the specification of the interface. This is shown in Figure 6.3 and defines an ATM layer and an ATM adaptation layer above the physical layer but below layers 2 and 3 of the OSI model. In addition, separate user, control and management planes are identified to reflect the possibility of different procedures in these planes.

---

2. This diagram appeared first in the Blue Book version of Recommendation I.121 whose later versions are a pale shadow of its former self. The recommendations produced since 1988 include much that was originally contained in Recommendation I.121.
3. RACE, for example, developed a reference model that identified seven layers covering OSI layers 1–3.

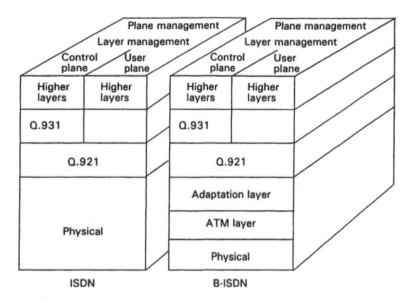

Figure 6.3 B-ISDN protocol reference model.

Figure 6.3 is based upon Figure 1 of Recommendation I.321[108] but shows also an equivalent representation of the ISDN architecture. Recommendation I.320[109], which is the equivalent document for ISDN, does not show the protocol reference model in the same way.

Figure 6.4 identifies the functions of the layers of the protocol reference model in a little more detail. These will be discussed in what follows.

	Higher layer functions	Higher layers	
	Convergence Segmentation and reassembly	CS SAR	AAL
	Generic flow control Cell header generation/extraction Cell VPI/VCI translation Cell multiplex and demultiplex		ATM
Layer management	Cell rate decoupling HEC header sequence generation/verification Cell delineation Transmission frame adaptation Transmission frame generation/recovery	TC	Physical layer
	Bit timing Physical medium	PM	

Figure 6.4 Functions of the B-ISDN in relation to the protocol reference model (Figure 2 of Recommendation I.321). CS, convergence sublayer; PM, physical medium; SAR, segmentation and reassembly sublayer; TC, transmission convergence.

137

6.3.1.1 ATM Physical layer

The definition of the physical layer described in ITU-T Recommendation I.321 makes provision for the use of either ATM or SDH as the transmission method. However, the format of the document and of Recommendation I.432[85]4 favours the use of ATM and quite a good deal of the structure is influenced by ATM considerations.

There are two sublayers in the physical layer. The physical medium (PM) sublayer contains only those functions that are dependent on the physical medium. The transmission convergence (TC) sublayer performs all the functions required to transform a flow of cells into a flow of data units (bits) to be transmitted and received over the physical medium. Thus, the service data unit (SDU) crossing the boundary between the ATM layer and the physical layer will be a flow of valid cells. The physical layer converts this into a data flow to be inserted into the transmission system payload, which is independent of the physical medium and is self-supported. To achieve this, the physical layer merges the ATM cell flow with appropriate information for cell delineation and operations and maintenance (OAM) information related to this cell flow.

The principle of the cell delineation algorithm is shown in Figure 6.5. The mechanism operates on the 32 bits of the first four octets of the header, checking that these match the header error control (HEC) octet, which has been

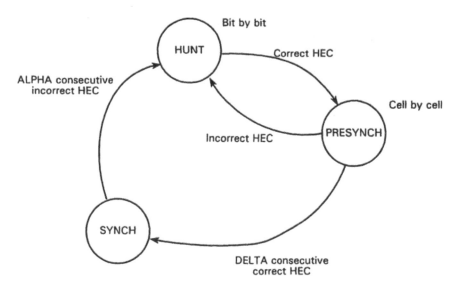

Figure 6.5 Cell delineation state diagram (Figure 5 of Recommendation I.432).

---

4. The current issue is dated 1993 but amendments dated November 1994 are considered here also. These amendments extend the applicability of the document to interfaces at 1.544 and 2.048 Mbit/s carrying ATM cells.

138

Table 6.2 Header pattern of OAM cell identification.

	Octet 1	Octet 2	Octet 3	Octet 4	Octet 5
Header pattern	00000000	00000000	00000000	00001001	HEC = valid code

derived from the first four octets using a shortened cyclic code with the generating polynomial $x^8 + x^2 + x + 1$. Table 6.2, which shows the header pattern for the OAM cell identification, helps to visualize the header arrangement.

The ATM cell stream is scrambled to increase the robustness of the delineation process against malicious use or unintended simulations of a correct HEC in the information field. The scrambling polynomial is defined for the SDH-based physical layer but not yet for the ATM-based physical layer. Only the information field is scrambled, the five header octets are sent as they are. The physical layer OAM cell and, possibly, other cells related to physical layer management must be identified from the pattern illustrated in Table 6.2 or from other values and extracted so that they are not passed to the ATM layer.

Other functions of the physical layer identified in Figure 6.4 need no great explanation. The function of transmission frame adaptation has little relevance where the payload structure of the transmission frame is cell equivalent (i.e. ATM, no external structure is added to the frame). In other cases, frame adaptation will be relevant where SDH or G.703[111] envelopes are in use. Cell rate decoupling inserts idle cells as necessary to adapt the rate of the ATM transmission to the payload capacity of the transmission system. The format of the idle cell header is shown in Table 6.3. Idle cells are not passed to the ATM layer.

Table 6.3 Header pattern for idle cell identification.

	Octet 1	Octet 2	Octet 3	Octet 4	Octet 5
Header pattern	00000000	00000000	00000000	00000001	HEC = valid code

6.3.1.2 ATM layer[88]

In discussing Figure 5.6, the virtual path identifier (VPI) and the virtual circuit identifier (VCI) were introduced and their function to identify the path over which the cell belonging to a call will be sent (VPI) and the call to which a cell belongs (VCI) was explained.

Cell multiplexing combines the various cells from individual virtual paths (VPs) and virtual circuits (VCs) into a non-continuous, composite flow of cells. Demultiplexing directs the cells to the appropriate VP or VC.

VPI and VCI translation provides the function of the ATM switch. The value of the VPI and/or VCI fields of each incoming ATM cell is mapped, where necessary, into a new VPI and/or VCI value.

Where the ATM layer is terminated and the messages are received from and sent to higher layers, cell header generation and extraction takes place. In the transmit direction the cell information field is received from a higher layer and the appropriate ATM cell header is generated, excluding the HEC sequence.

This operation might, for example, include translation from a service access point (SAP) identifier (ISDN DSS 1 layer 2 protocol[34]) to a VP and VC identifier. In the receive direction, the cell header extraction function removes the ATM cell header and passes the cell information field to a higher layer, converting, where necessary, the VP and VC identifier into an SAP identifier.

The ATM layer termination may well be a B–NT1 or B–NT2 interface to several user devices. In this case, generic flow control (GFC) may be applicable and the cells carrying this information to control the point-to-multipoint connections will be generated in the ATM layer.

Communication between the ATM layer and higher layer is via the ATM adaptation layer immediately above it. This communication comprises the service data unit (SDU), that is the ATM information field, and, at least, the following primitives:

    (a) ATM-DATA-request: AAL requests the ATM layer that the ATM-SDU associated with this primitive be transported to its peer.

    (b) ATM-DATA-indication: the AAL is notified by the ATM layer that the ATM-SDU associated with the primitive coming from its peer is available.

6.3.1.3 ATM *adaptation layer*[89, 90]

A good principle in dealing with problems is to collect them all together in one place. The ATM adaptation layer successfully does this for the ATM-based B-ISDN. As its name implies, the adaptation layer is intended to adapt the ATM cell-based information coming from the ATM layer to the various possible user protocols in the layers above, which will be defined with no consideration whatsoever for the exigencies of ATM. As an example, in the case of services already offered by the ISDN passing via ATM, the layer above the ATM adaptation layer will be DSS 1 signalling data link layer[34].

Thus, the AAL layer isolates the higher layers from the specific characteristics of the ATM layer by mapping the higher layer protocol data units to the information field of the ATM cell and vice versa. To achieve this function the AAL is organized into two logical sublayers as follows:

    (a) SAR (segmentation and reassembly). The prime functions are segmentation of higher layer information into a size suitable for the information field of the ATM cell and reassembly of the contents of the ATM cell information fields into higher layer information.

    (b) CS (convergence). The prime function is to provide the AAL service at the AAL-SAP. In other words, this sublayer is service dependent.

In theory, therefore, there could be as many different AALs as there are service protocols at the higher levels to be accommodated by ATM. In each case the CS sublayer at least could be different. To attempt to avoid this tremen-

Table 6.4 Service classification for AAL (Figure 1 of Recommendation I.362).

	Class A	Class B	Class C	Class D
Timing relation between source and destination	Required		Not required	
Bit rate	Constant	Variable		
Connection mode	Connection-oriented			Connectionless

dous proliferation of different AALs Recommendation I.362 attempts to define a classification of AALs in terms of service using the following parameters to aid the definition: whether there is a need for a timing relation between source and destination; whether the bit rate is constant or variable; and whether a connection-oriented or connectionless mode of operation is required. Using these parameters I.362 foresees four classes of AAL as shown in Table 6.4.

Recommendation I.363 uses this categorization to define a number of AAL types. As they stand, the recommendations make it very clear that not all the possibilities are covered and that there are possibilities where the AAL will be empty, reduced to the reception and delivery of ATM-SDUs. At the present time, Recommendation I.363 defines the following types of AAL:

(a) AAL type 1: to provide constant bit rate services.
(b) AAL type 2: to provide variable bit rate services, presumably those not provided by AAL type 3/4. The definition of AAL type 2 in Recommendation I.363 is not complete.
(c) AAL type 3/4: to provide connectionless service (class D) and to provide class C services (where timing relationship is not required) and for signalling.

The layer services provided by AAL type 1 to the AAL user are: transfer of service data units with a constant source bit rate and the delivery of them with the same bit rate; transfer of timing information between source and destination; transfer of structure information between source and destination; and indication of lost or errored information that is not recovered by AAL type 1, if needed.

The layer services provided by AAL type 2 to the AAL user may include: transfer of service data units with a variable source bit rate; transfer of timing information between source and destination; and indication of lost or errored information that is not recovered by AAL type 2, if needed.

Both AAL type 1 and AAL type 2 are envisaged as having a structure of two sublayers: SAR and CS. For AAL type 1 the definitions are fairly complete, but for type 2 there are still many gaps.

Historically the four classes of Table 6.4 were seen as requiring four types of AAL. As definition progressed, however, the functions of AAL type 3 and AAL type 4 were seen as convergent and there is now a single definition of AAL type 3/4.

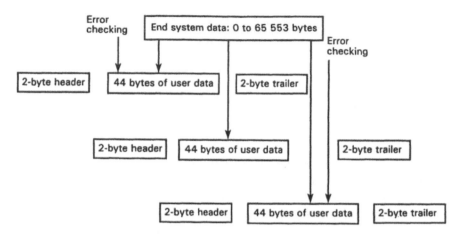

Figure 6.6 Overview of the function of AAL type 3/4.

AAL type 3/4 is intended to transfer variable bit rate, bursty data that can tolerate delay but not loss. It is the most complex and the most completely defined of the AAL types described in Recommendation I.363. To attempt to see first an overview of the objective, Figure 6.6 attempts to summarize what is happening. AAL type 3/4 is defined differently from AALs types 1 and 2 and is envisaged as having a different structure, shown in Figure 6.7.

Two modes of service are defined for AAL type 3/4: message and streaming. In the *message-mode service* the AAL service data unit is passed across the AAL interface in exactly one AAL interface data unit (AAL-IDU). This service provides the transport of fixed size or variable length AAL-SDUs.

(a) In the case of small fixed-size AAL-SDUs an internal blocking and deblocking mechanism in the SSCS may be applied. This provides for the transport of one or more fixed-size AAL-SDUs in one SSCS-PDU.

(b) In the case of variable length AAL-SDUs an internal AAL-SDU message segmentation/reassembling function in the SSCS (Figure 6.7 explains these acronyms and abbreviations) may be applied. In this case, a single AAL-SDU is transferred in one or more SSCS-PDUs.

(c) Where these options are not used, a single AAL-SDU is transferred in one SSCS-PDU. When the SSCS is null, the AAL-SDU is mapped to one CPCS-SDU.

In the *streaming-mode service* the AAL-SDU is passed across the AAL interface in one or more AAL-SDUs. The transfer of these AAL-SDUs across the AAL interface may occur with separations in time. This service provides the transport of variable-length AAL-SDUs. The streaming-mode service includes an abort service by which the discarding of an AAL-SDU partially transferred across the AAL interface can be requested.

(a) An internal AAL-SDU message segmentation/reassembly function in the SSCS may be applied. In this case, all the AAL-IDUs belonging to a single

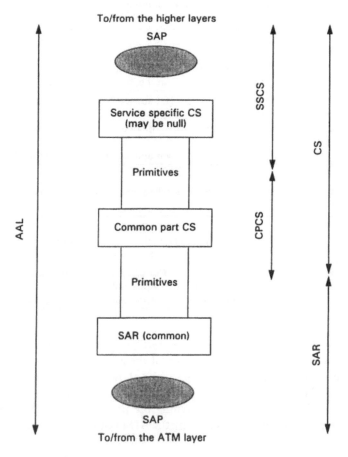

Figure 6.7 Structure of AAL type 3/4 (Figure 10 of Recommendation I.363). CS, convergence sublayer; CPCS, common part convergence sublayer; SAP, service access point; SAR, segmentation and reassembly sublayer; SSCS, service-specific convergence sublayer.

AAL-SDU are transferred in one or more than one SSCS-PDU.
(b) An internal pipelining function may be applied. This provides the means by which the sending AAL entity initiates the transfer to the receiving AAL entity before it has the complete AAL-SDU available.
(c) Where option (a) above is not used, all the AAL-IDUs belonging to a single AAL-SDU are transferred in one SSCS-PDU. When the SSCS is null, the AAL-IDUs belonging to a single AAL-SDU are mapped to one CPCS-SDU.
Recommendation I.363 attempts to summarize these descriptions with the tabulations reproduced in Tables 6.5 and 6.6.

The methods of working that these definitions permit are shown in Figure 6.8. This is taken from one of the appendices of Recommendation I.363,

Table 6.5 Combination of service mode and internal function, AAL type 3/4 (Table 3 of Recommendation I.363).

	AAL-SDU message segmentation/ reassembly in the SSCS	AAL/SDU blocking deblocking in the SSCS	Pipe lining
Message option 1 Long variable size SDUs	Optional	Not applicable	Not applicable
Message option 2 Short fixed size SDUs	Not applicable	Optional	Not applicable
Streaming	Optional	Not applicable	Optional

Table 6.6 Combination of service mode at the sending and receiving side (Table 4 of Recommendation I.363).

	Sender		
	Message mode		Streaming mode
Receiver	Block	Segment	
Message-mode deblocking	Applicable	Not applicable	Not applicable
Message-mode reassembly	Not applicable	Applicable	Applicable
Streaming mode	Not applicable	Applicable	Applicable

which are well worth looking at first, before trying to come to grips with the detail of the protocol given in the main body of the recommendation.

It is indicated in Recommendation I.363 that AAL type 5 is "under study". This illustrates well the fact that there are many different organizations working on ATM, not all of which are approaching the problems from the same direction. AAL type 5 has been conceived by the ATM Forum and adopted, in

(a)

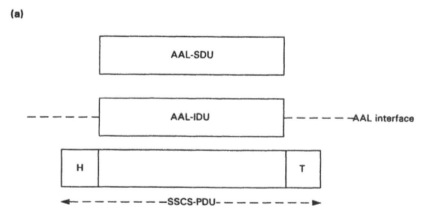

Figure 6.8 Types of service at the AAL type 3/4 interface (Figure A.3 of Recommendation I.363). (a) Message-mode service.

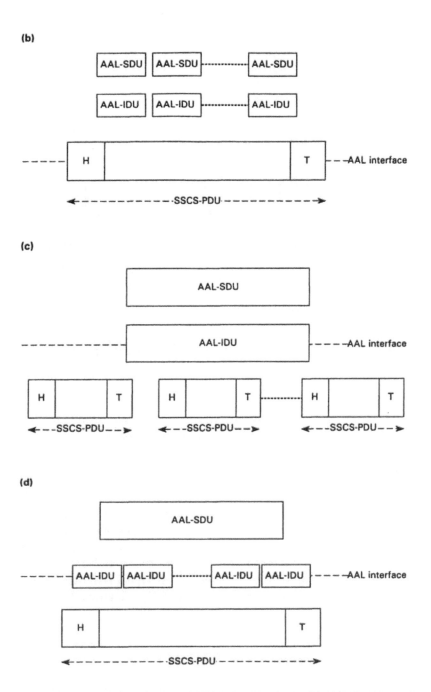

Figure 6.8 Types of service at the AAL type 3/4 interface (Figure A.3 of Recommendation I.363). (b) Message-mode service plus blocking/deblocking internal function. (c) Message-mode service plus segmentation/reassembly internal function. (d) Streaming-mode service.

145

(e)

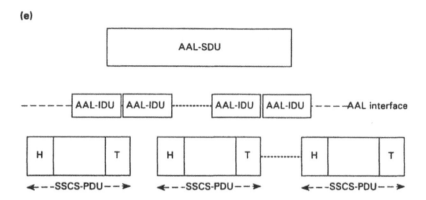

Figure 6.8 Types of service at the AAL type 3/4 interface (Figure A.3 of Recommendation I.363). (e) Streaming-mode service plus segmentation/reassembly internal function.

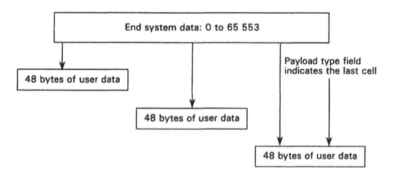

Figure 6.9 Overview of the function of AAL type 5.

principle, by ANSI and the ITU as the simple and efficient adaptation layer (SEAL). Supporting only message mode, non-assured types of operation, it is conceived as a subset of AAL type 3/4. The SSCS is null and the CPCS-PDU is divided into 48-byte payloads lacking the two-byte headers and trailers of the SSCS-PDU illustrated in Figure 6.8. The effect of AAL type 5 is illustrated, in a similar fashion to that of Figure 6.6, for AAL type 3/4, in Figure 6.9.

6.4 ATM services, another view, the ATM Forum

One of the driving forces defining ATM is the ATM Forum in the USA. This has defined various types of ATM service equivalent but not analogous to the types of service defined in connection with the AAL adaptation layer in the previous section. It is likely that the insights of the ATM Forum will influence future

146

definition by ITU-T so that this other view has relevance to our discussion even though it may complicate the issues.

The ATM Forum has defined five classes of ATM service that the ATM network might provide.[5] These are:

(a) CBR (continuous bit rate) service is a deterministic service to support circuit emulation, voice traffic and continuous bit rate video. It provides reserved bandwidth with minimal cell loss or variation in delay.

(b) VBR (variable bit rate) is defined as two types as follows: VBR-RT (real time) provides tight bounds on delay for transmission of voice, and video with silent periods removed; VBR-NRT (non-real time) has less stringent bounds on delay and was developed to provide a class of service for transmission of transaction data.

(c) UBR (unspecified bit rate) provides best-effort delivery but offers no guarantee on cell loss or delay variation. It is designed to allow use of excess bandwidth not consumed by the other services that use a feedback loop to adjust their transmissions to the quality of the received messages.

(d) ABR (available bit rate) also exploits excess network bandwidth, but in this case closed-loop feedback is used to allow end stations to reduce their transmissions to avoid cell loss. This service is intended for transmitting messages between LANs and other bursty, unpredictable data traffic.

The feedback mechanisms use a number of measurements, known as service descriptors, to ensure the quality of service required by the particular class of service. These are:

(a) PCR (peak cell rate).

(b) MCR (minimum cell rate): the minimum cell rate guaranteed by the service provider.

(c) CDVT (cell delay variation tolerance): the tolerance for the difference in delay between the cell that is delayed the least and the cell with the greatest delay.

(d) SCR (sustained cell rate): the average rate of transmission of cells sustained over the duration of the transmission.

(e) BT (burst tolerance): the limit (in time) that the transmission can maintain its peak rate (PCR).

An ATM service might be expected therefore to guarantee in the contract for service some or all of these parameters. Table 6.7 shows the view of the ATM Forum on the content of the service guarantees for all the defined ATM classes of service.

Although there is some coincidence with the classes of service indicated by the ITU-T recommendations for the adaptation layer, it can be seen from the deliberations of the ATM Forum that the definition process is by no means complete.

---

5. Much of this material is drawn from reference 112.

Table 6.7 ATM classes of service and their guarantees.

| Service | Descriptors | Guarantees | | | |
		Loss	Delay	Bandwidth	Feedback
CBR	PCR, CCDV	Yes	Yes	Yes	No
VBR-RT	PCR, CCDV, SCR, BT	Yes	Yes	Yes	No
VBR-NRT	PCR, CCDV, SCR, BT	Yes	Yes	Yes	No
UBR	Unspecified	No	No	No	No
ABR	PCR, CCDV, MCR	Yes	No	Yes	Yes

6.5 ATM operation and maintenance provisions

A hierarchy of levels of possible faults has been defined for ATM. This hierarchy is illustrated in Figure 6.10.

Each fault indication can be propagated in both directions from the source of the trouble and will be carried in the relevant cells or messages of the transmission protocol in use. Thus, assuming that an SDH transmission system is in use, the meaning of the fault indications will be as follows:

(a) F1, F2: loss of frame – SDH frame synchronization has been lost; or degraded error performance – the quality of the received bit stream is not acceptable.

(b) F3: loss of cell delineation; uncorrectable header – the header has more errors than can be corrected; degraded header error performance; loss of AU4 pointer;[6] degraded error performance; failure of the insertion and suppression of idle cells.

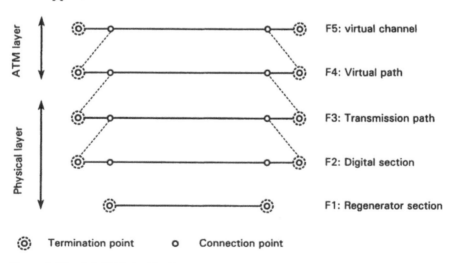

Figure 6.10 O & M hierarchical levels.

---

6. See the discussion of SDH in Chapter 7.

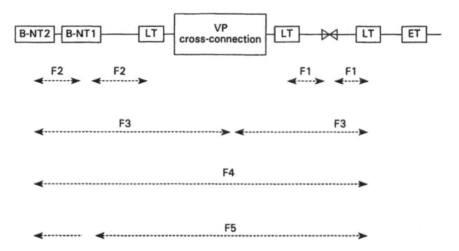

Figure 6.11 OAM flows in a typical configuration.

On the other hand, were this to be a cell-based (ATM) transmission system then the meaning of the fault categories could be as follows:

(a) F1, F2: loss of physical layer OAM (PLOAM) cell recognition.

(b) F2: degraded error performance.

(c) F3: loss of cell delineation; uncorrectable header; degraded header error performance; failure of the insertion and suppression of idle cells.

For a typical physical configuration the flows of these error indications in terms of source and destination are shown in Figure 6.11. At the ATM layer the OAM messages provide the following indications:

(a) F4: path not available.

(b) F4, F5: degraded performance.

Chapter summary

The chapter began with a short overview of the development story of the B-ISDN with particular reference to the development of ATM. This led to the concept of a fixed small cell or packet being used to transmit all information, voice or data, and the rather untidy compromise between the data people, who wanted a cell of a useful size, and the voice people, particularly the Europeans, who feared that a large cell would involve a much greater use of echo control.

Having fixed the size of the cell to nobody's satisfaction, the definition of the user–network interface and the protocol to be used to transmit ATM were described. It was seen that basically the protocol inserts some extra layers between the transmission line and the ISDN layers 2 and 3 defined for DSS 1.

It was shown that the majority of the difficulty and complication of the protocol was concentrated in the adaptation layer and the versions of this layer that are presently defined were discussed, not failing to note that there may be a requirement for additional definitions. This view was supported by reference to the views of the ATM Forum.

The discussion concluded with a short description of the operations and maintenance provisions of the ATM standards as presently defined.

CHAPTER 7

Broadband ISDN tools: synchronous digital hierarchy (SDH)

Chapter 2 included a diagram, Figure 2.13, showing the PCM coding technique in the manner that has become accepted, as a long strip. By redrawing this as shown in Figure 7.1, it is apparent we have all along been using the concept underlying SDH.

Column	1	2	3	4	5	6	7	8	9	10	11	12	13	14	15	16
Row 1	0	1	2	3	4	5	6	7	8	9	10	11	12	13	14	15
Row 2	16	17	18	19	20	21	22	23	24	25	26	27	28	29	30	31
	Overhead															

Figure 7.1 PCM coding techniques[19].

The overhead can and does contain indications of where the remainder of the information in the envelope can be found. Channel 16 can contain the signalling information relating to any of the other channels in the envelope. Of course, it is known where everything is because the transmission is synchronous, the timing is tracked and any channel can be accessed directly. However, illustrated like this, it can be seen that the envelope did not have to contain equal-sized pieces; it could have transmitted wide-band information using several channels, not necessarily adjacent, and pointed to them in the overhead. This is exactly what SDH does on a much larger scale.

Public transmission network planning for the next few decades is, as we have seen, predicated upon the massive introduction and increase of optical transmission in the network. The plans assume that this high-bandwidth optical network will be managed and controlled through the use of the Synchronous Digital Hierarchy (SDH). A diagram illustrating the UK SDH network plan now being put in place is shown at Figure 7.2a, and other national plans presently formulated are not dissimilar to this.

The network configurations illustrated in Figure 7.2a were first concieved several years ago. More modern versions of similar networks would envisage more detailed methods of safeguarding the important resources of the main

(a)

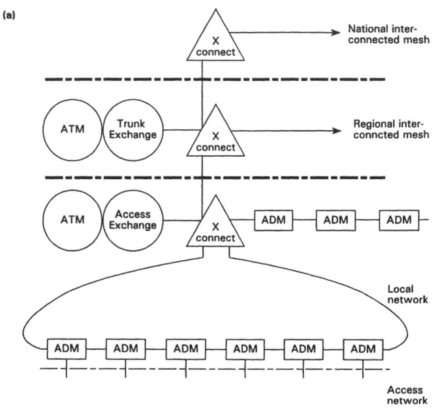

Access networks such as: copper   Connecting to: subscribers, large and small
                          fibre   remote concentrator and switching units
                          passive optical   mobile radio networks
                          networks   other public networks

(b)

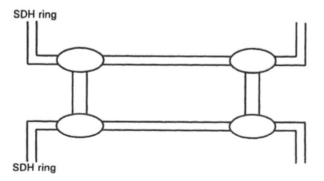

Figure 7.2   (a) UK transmission network plan. (b) Secure interconnection of SDH rings, bi-directional line switched rings.

ring structures circulating information at 155 Mbit/s, 620 Mbit/s and higher data rates. One such improvement is illustrated in Figure 7.2b. Here, the point of intersection between two such rings is shown. Instead of interconnecting via a single node, two interconnection nodes would be allocated, each interconnected by different trunks in different routes. Failure of any trunk or any node is therefore covered by the other without any need for switchover on failure. Such arrangements are already in use in national networks. One such, known to the author, is in Vietnam.

7.1 SDH in principle

The story of SDH arises out of the development of leased circuit networks, particularly in the US. It has already been told in Chapter 5.

In writing this book the author has had to assemble a large amount of information from books, standards, articles from journals, etc. He has had to collate this information and produce contents lists and indices in order to be able to access the passages he needs when he comes to the relevant subject. In the same way, the reader will be less than delighted with the book unless it provides indices and glossaries as a guide to the parts that need to be read for any particular purpose.

The principle of SDH is not unlike this. Lump together a large amount of information from a variety of sources and of varying sizes and put it all in the one envelope. Send along with it a catalogue of the information contained, giving precise information of where it is to be found. This is SDH.

To transmit the 32 8-bit bytes of Figure 7.1 8000 times a second (in order to meet the Nyquist criterion for validity of sampled analogue information[1]) requires a bandwidth of:

$$16 \times 2 \times 2 \times 8000 = 2.048 \, \text{Mbit/s}$$

With similar reasoning, transmission at about 50 Mbit/s, chosen for the same reason as in the ATM argument, to accommodate a TV channel supposed to require some 44 Mbit/s:

$$90 \times 9 \times 8 \times 8000 = 51.84 \, \text{Mbit/s}$$

This gives an envelope that can be visualized as 90 bytes wide and 9 bytes deep in which to contain the information to be transmitted. This is in fact the envelope chosen for SONET, the American SDH system from which SDH itself was derived.[2] For SDH it was decided that a more flexible and usable basic module should be three times this size: an envelope 270 bytes wide and 9 bytes deep.

$$270 \times 9 \times 8 \times 8000 = 155.520 \, \text{Mbit/s}$$

---

1. See any digital transmission text book. Reference 2 covers the subject.
2. A good account of the CCITT/ANSI compromise is contained in reference 113.

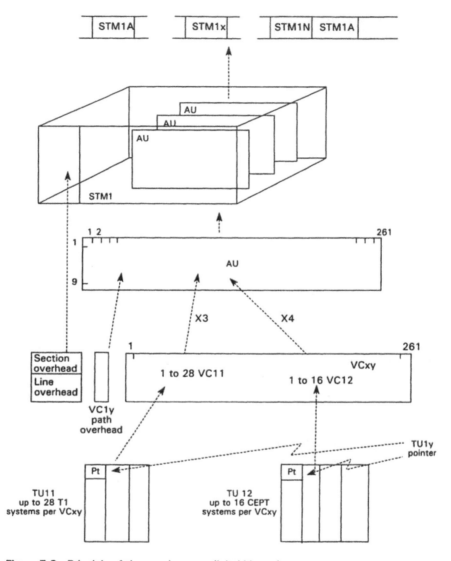

Figure 7.3 Principle of the synchronous digital hierarchy.

How this envelope is used is shown in Figure 7.3.

Figure 7.3 shows as examples the mapping of North American 24-channel PCM systems (T1) to ITU-T Recommendation G.733[114] and 32-channel systems (CEPT) to ITU-T Recommendation G.732[19]. For 24 channels, three columns of the STM envelope are allocated as a tributary unit (TU). These 27 bytes provide space for 1 byte overhead as a pointer and the 24 bytes carrying the channel bytes of the PCM frame. Similarly, the 32-channel system is allocated a TU-12 consisting of four columns of the STM envelope, giving room for a pointer byte and 32 bytes carrying the channels of the PCM frame.

In the complete envelope certain portions are defined such that they always occur in a fixed position. These are the first nine columns of the envelope and an additional column for every assembly of similar TUs, called a virtual container (VC). This is the overhead, or most of it; additional elements will be introduced in what follows. With the envelope transmitted synchronously it will always be possible to find these rows just by counting from the start of the frame. They are fixed in just the same way as channels 0 and 16 are fixed in the PCM envelope. However, whereas in PCM every channel is fixed, here nothing else is fixed. The tributary units can go anywhere – their actual position is determined from a pointer contained in the overhead.

So, tributary units can go anywhere and, theoretically, in any mixture of different tributary units. The diagram implies that they are all TU-11 or all TU-12, but this need not be the case. Having assembled a number of TUs to fill the 261 columns of available space, this is combined with the overheads to make up an administrative unit (AU). Were this SONET, this would be the end of the matter, but with three times the bandwidth available in SDH we can assemble a number of AUs into the so-called synchronous transfer module 1 (STM-1), the basic transmission unit of the SDH system.

7.2 SDH structure

The basic building block of the SDH is visualized as a pattern of 270×9 bytes each of 8 bits repeated once every 125 µs. Thus, each byte provides a 64 kbit/s channel and the complete pattern, STM-1,[3] provides a communications capacity of 155.520 Mbit/s (in SONET 51.840 Mbit/s). STM-1s can be remultiplexed together for transmission at the higher levels, as shown in Table 7.1. Alternatively, AUs can be assembled into higher order envelopes, STMNs, as explained in subsection 7.2.2. Figure 7.3 also illustrates the structure.

Table 7.1 also clarifies some of the confusing differences of terminology between SDH and SONET. As they are both current systems in use it is perhaps important to have a clear idea of the differences. Within the SDH STM the first nine columns (81 bytes) are reserved for overhead and the remaining space, in SONET known as the payload envelope, is an AU.

Of the management signalling in the overhead bytes, 27 bytes are reserved for repeater section overhead and line overhead and would be common for all the contents of the STM-1. A further 9 bytes is reserved for path overhead and would be common for a particular type of virtual container (VC) being carried by the STM. Each tributary unit (TU) is mapped into a corresponding VC together with a TU pointer within the remaining 261×9 bytes of the AU. Figure 7.3 illustrates this process for TU-11 (1.544 Mbit/s T1 channels allocated a

---

3. In SONET the equivalent is a synchronous transport signal level 1 (STS-1) but is dimensioned as 90×9 bytes of 8 bits.

Table 7.1 The synchronous digital hierarchy.

SDH term	Tributary unit, TU *Maps onto:*	Virtual container, VC *Combines into:*	Administrative unit, AU *Combines into:*	Synchronous transport module module, STM	
SONET term	Tributary	Virtual tributary, VT	Synchronous transport signal, STS*	STS 3 or OC 3[4]	
				Equivalent tributary rate (Mbit/s)	Number of SDH envelope columns
	TU 11	VC 11	AU 11	1.544	3
	TU 12	VC 12	AU 12	2.048	4
	TU 21	VC 21	AU 21	6.312	12
	TU 22	VC 22	AU 22	8.448	16
	TU 31	VC 31	AU 31	34.368	65
	TU 32	VC 32	AU 32	44.736	85
	TU 4	VC 4	AU 4	139.264	261

*In SONET, STS-1 consists of data communications channels: transport overhead consisting of section overhead and line overhead and occupying three columns and synchronous payload envelope consisting of path overhead plus the VCs, occupying 87 columns.

VC-11 of 3×9 bytes) and for TU-12 (2.048 Mbit/s CEPT channels allocated a VC-12 of 4×9 bytes). Thus, the STM can carry 84 T1 or 64 CEPT channels or a mixture. The contents of a VC are clearly not limited to digitally encoded voice but will include packet data, circuit-switched data, video, etc.

The nomenclature for containers, VCs, TUs, and tributary unit groups (TUGs) uses two succeeding digits x,y, to specify the multiplexing level, x, and the bit rate option, y, respectively (see Table 7.1).

Figure 7.3 shows a mapping of TU-11s and TU-12s onto the STM and suggests that all TUs are of the same kind. Although this may be the case in the majority of applications, SDH makes provision for the mixing of different kinds of TU. For each kind however, there must be a separate VC-1y path overhead. Thus, an AU could contain, for example, 37 TU-11s [111 columns for 37 VC-11s (T1) plus one column VC-11 path overhead] and 37 TU-12s [CEPT; 148 columns plus one column VC-12 path overhead]. This entirely fills the 261 columns available in the STM. It may be, however that such an arrangement is not permitted by the particular subset of the recommendations adopted in particular parts of the world. Although Table 7.1 shows the column count for higher order VCs, these may vary in practice, particularly at the VC-2 level, depending on the method of mapping chosen.[5]

---

4. The alternative SONET term is optical carrier (OC).

5. The BT multiplexer specification RC 8495 makes provision for the mixing of VC types within an AU but does indicate that this may prove to be unusual in practice.

7.2.1 Tributary mapping

Figure 7.3 indicates the presence of a tributary unit pointer but does not demonstrate its purpose as it is shown as part of the virtual container to which it relates. The provision of the pointer, however, allows freedom of choice of mapping algorithm to be used and freedom for the SDH to accept a flexible construction of the administrative units.

As an introduction, Figure 7.4 illustrates, out of many, one form of mapping for a 34.368 Mbit/s payload. This and much of the material of this chapter is taken from the ITU-T recommendations for the SDH[96, 97, 163].

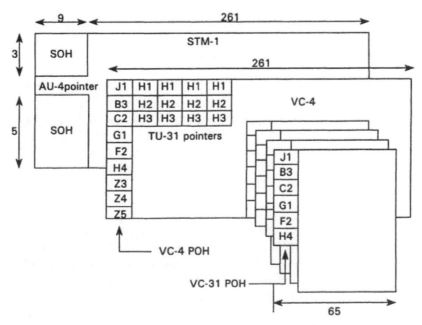

Figure 7.4 Mapping of four VC-31s into an AU-4 (Figure 2-2 of Recommendation G. 709, 1988).[6]

This diagram shows how a container, C-31, a 34.368 Mbit/s signal, becomes a virtual container, VC-31, through the addition of its pointers, H1, H2 and H3, which indicate where inside the virtual container each of the four containers VC-31 making up the complete VC-4 container may be found. The VC-4, in turn, becomes an administrative unit, AU-4, through the addition of the AU-4 pointer, located in the first 9 bytes of row 4 within the path overhead

---

6. This figure is from the Blue Book, 1988, version of G.709 [97]. The 1993 version is very different and does not contain this particular diagram at all. Section 7.2.3 explains the history of the development of SDH from 1988 to 1993. The old diagram is retained because the author feels that it is more helpful in illustrating the explanatory text.

of the complete STM-1 envelope, indicating where inside the STM-1 the administrative unit may be found.

As the synchronous signal varies in frequency over time, the transmitting and receiving equipments recognize and synchronize themselves to the frame alignment information in just the same way as does the ordinary primary rate PCM system. Beyond this the pointer mechanism is being continuously updated so that the particular portion of the envelope containing a particular communications channel is indicated precisely by the pointer. The receiving and transmitting equipment can, where necessary, identify this portion just by counting from the known position of the beginning of the frame according to the value of the pointer.

7.2.2 The development of the SDH structure

Figure 7.4 (from Recommendation G.709, 1988) was included to help with the explanation. By 1993 this figure had disappeared from the recommendation and the context of the explanations and the descriptions had changed. The basic difference is that in 1988 an SDH envelope was considered only at the STM-1 level and conventional bit interleaved multiplexing above that was assumed whereas, by 1993, the envelope was described at the STM-N level, where N can be 1 (155.520 Mbit/s), 4 (622.080 Mbit/s) or 16 (2.488 Gbit/s).

In 1988 the agreement envisaged in moving from SONET to SDH was that the SONET envelope at 51.840 Mbit/s would be mapped into SDH as three SONET STS-1 envelopes. Annex A to Recommendation G.708 defines this and indicates that two columns of fixed stuff are used to replace the redundant columns of path overhead from two of the three SONET envelopes.

If this can be done to SONET envelopes to include them in SDH then, instead of going to higher order SDH multiplexes by bit interleaving and therefore requiring demultiplexing to access the STM-1 envelope, it is just as easy to consider higher order envelopes. Thus, the 1993 recommendations describe the arrangements in terms of STM-N rather than STM-1 and have introduced the concept of administrative unit groups (AUGs) into envelopes at an order higher than 155.520 Mbit/s.

To illustrate this development the multiplexing structure envisaged in 1988 is reproduced in Figure 7.5 and the current arrangement in Figure 7.6. These figures also illustrate that there are many more multiplexing and mapping opportunities than are recommended to be implemented. In practice, network operators can choose a subset of those shown in Figure 7.6 for implementation.

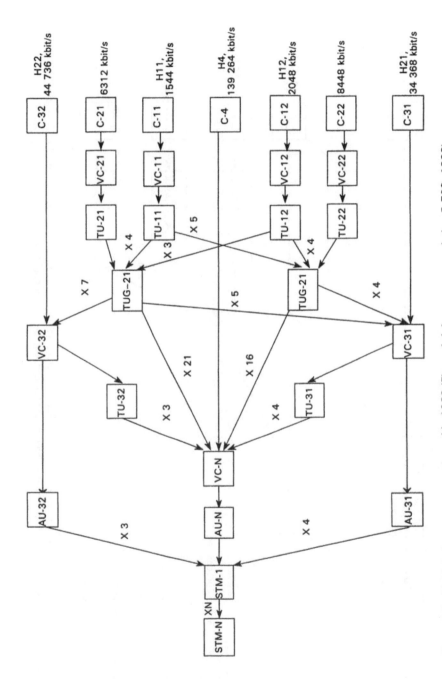

Figure 7.5 Multiplexing structure envisaged in 1988 (Figure 1-1 from recommendation G.709, 1988).

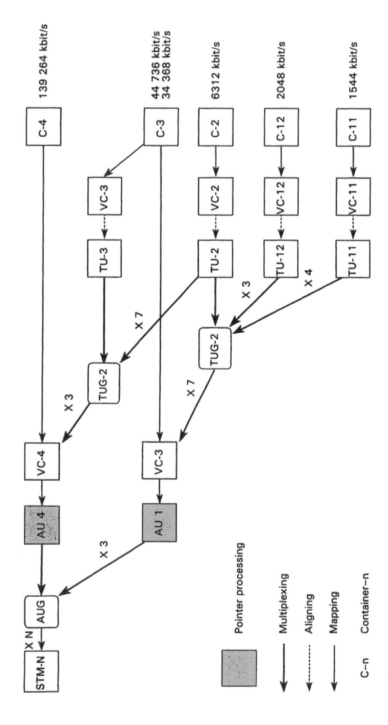

Figure 7.6 Multiplexing structure envisaged in 1993 (Figure 1-1 from recommendation G.709, 1993). Note that G.702[115] tributaries associated with container C-x are shown. Other signals, e.g. ATM, can also be accommodated.

7.2.3 Multiplexing, mapping and alignment

Figure 7.6 introduces the triple concepts of multiplexing, mapping and aligning. The distinction can best be illustrated by showing the generalized method whereby containers are assembled into the STM-N signal. Figure 7.7 illustrates this using the C-11 of Figure 7.6 as an example. G.708 has a similar set of pictures illustrating the process that are entirely generalized (Figures 2-1 to 2-5 of Recommendation G.708[163]).

The original 1.544 Mbit/s, 24 -channel PCM signal is the container, C-11, of the information. This is *mapped* to the virtual container, VC-11, by adding the VC-11 path overhead. Putting these two together and constructing the TU-11 pointer indicating where they may be found is the process of *aligning* to

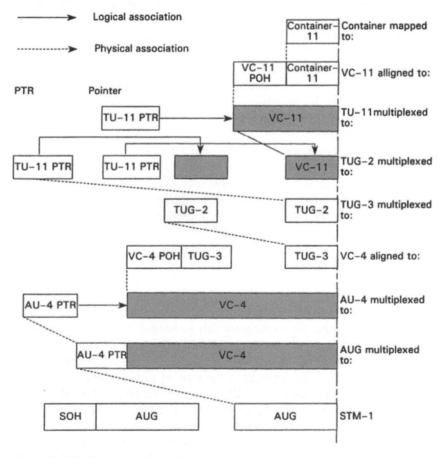

Figure 7.7 Multiplexing method directly from container C-11 using AU-4 (based on Figure 2-2 of Recommendation G.708). Note that unshaded areas are phase aligned. Phase alignment between the unshaded and shaded areas is defined by the pointer (PTR) and is indicated by the arrow.

the tributary unit TU-11. This TU-11, pointer and virtual container are in turn *multiplexed* with other tributary units into the tributary unit group TUG-2, which, as a glance back at Figure 7.6 will remind us, could be made up of all C-11s or a mixture of C-11s, C-12s or C-2s. Now a number of TUG-2s can be multiplexed into a TUG-3 and the necessary VC-4 path overhead constructed. Putting these together and constructing the AU-4 pointer is the process of aligning into an AU-4, which is in turn multiplexed, pointer and VC-4, into the AUG, which is in turn multiplexed into the STM, where the section overhead is added.

Recommendation G.708 gives definitions of each of these processes as they are understood in SDH terminology. These are given below with comments relevant to the description we have just used.

(a) SDH mapping: a procedure by which tributaries are adapted into virtual containers at the boundary of an SDH network.

Mapping has been seen as putting the container bit by bit within the SDH structure.

(b) SDH aligning: a procedure by which the frame offset information is incorporated into the tributary unit or the administrative unit when adapting to the frame reference of the supporting layer.

This has been seen as the construction of and addition to the tributary of the pointer information.

(c) SDH multiplexing: a procedure by which the multiple lower order path layer signals are adapted into the higher order path or the multiple higher order path layer signals are adapted into a multiplex section.

The tributary units have been seen being multiplexed together and the administrative units have been seen being multiplexed into the STM-N.

7.2.4 SDH overhead

Everything discussed so far indicates that the breakthrough in SDH is the pointer mechanism that makes it possible to identify the part of the transmission of interest within the asynchronously presented contents of a synchronous envelope. It is now almost time to describe this in more detail, but before then the other elements of the SDH overhead will be briefly considered.

SONET allocates three columns at the start of the envelope to overhead, giving 3×9 bytes of information.[7] Of these, row 4 is reserved for the pointer information associated with the single AU that SONET is able to carry. SDH, with three times the capacity in an STM-1, allocates nine columns (81 bytes) to overhead so that there is capacity for up to three AUs. The pointer bytes for these AUs will again be in row 4, and the three bytes associated with the pointer of the first AU will be in the first three columns, the second AU pointer

7. The SONET terms for the overheads are section overhead (in SDH RSOH), the line overhead (in SDH MSOH) and the path overhead (as in SDH).

	9 bytes wide									
9 rows deep	A1	A1	A1	A2	A2	A2	C1	X̲	X̲	RSOH
	B1	D	D	E1	D		F1	X	X	
	D1	D	D	D2	D		D3			
	AU-n pointers									POH (part)
	B2	B2	B2	K1			K2			
	D4			D5			D6			MSOH
	D7			D8			D9			
	D10			D11			D12			
	Z1	Z1	Z1	Z2	Z2	Z2	E2	X	X	

X, Bytes reserved for national use. X̲, Underlining indicates unscrambled bytes, Therefore care should be taken in their content. D, Media dependent bytes.
Note: All unmarked bytes are reserved for future international standardization (for media dependent uses, additional national use and other purposes).

Figure 7.8 STM-1 section overhead (SOH) (Figure 5-2 of Recommendation G.708).

in the second three bytes and so on. The allocation of the remainder of the overhead is associated with the relevant AU in the same way. Similarly, for an STM-N there will be $N \times 9$ columns in the overhead and the overhead belonging to each AU will be in adjoining columns.

The location of the SOH byte in an STM-N frame is therefore identified by a three co-ordinate vector $S(a,b,c)$ where a (1–3, 5–9) represents the row number, b (1–9) represents a multicolumn number identifying a group of adjacent columns and c (1 –N) represents the depth of the interleaving within the multicolumn. For example, the second AU pointer in an STM-1 is located at $S(4,4,1)$.

This is all by way of explanation of Figure 7.8, which shows the arrangement of the various overhead indications but ignores the fact that for SONET there are just three columns, for SDH STM-1 each box is a byte but that for SDH STM-N each box is Nbytes wide by one column deep.

Figure 7.8 thus shows the identity of all the elements of the section overhead (SOH) and the use of these elements will be briefly discussed in what follows.

Figure 7.8 also shows the overhead divided into the regenerator section overhead (RSOH) and the multiplex section overhead (MSOH). The function of the different bytes is described using the text of Recommendation G.708, as follows:

(a) Framing: A1(11110110), A2 (00101000).
(b) STM identifier: C1. This is a unique identifier indicating the binary value of the multicolumn interleave depth co-ordinate "c".
(c) Data communications channel (DCC): D1 to D12. A 192kbit/s channel is

defined by bytes D1, D2 and D3 as a regenerator section DCC. A 576 kbit/s channel is defined by bytes D4 to D12 as a multiplex section DCC.

(d) Order wire: E1, E2. These two bytes may be used to provide order wire channels for voice communication. E1 is part of the RSOH and may be accessed at regenerators; E2 is part of the MSOH and may be accessed at multiplex section terminations.

(e) User channel: F1. This byte is reserved for user purposes (e.g. to provide temporary data/voice channel connections for special maintenance purposes).

(f) BIP-8: B1. One byte is allocated for regenerator section error monitoring. This function shall be a Bit interleaved parity 8 (BIP-8) code using even parity. The BIP-8 is computed over all bits of the previous STM-N frame after scrambling and is placed in byte B1 before scrambling.

(g) BIP-N × 24: B2. The three bytes[8] are allocated for a multiplex section error monitoring function similar to that described above.

(h) Automatic protection switching (APS) channel: K1, K2. Two bytes are allocated for APS signalling.

(i) Synchronization status: Z1 (b5 to b8). Bits 5–8 of Z1 in position S(9,1,1) are allocated for synchronization status messages. Four bit patterns are assigned to the four synchronization levels agreed within the ITU-T. Two additional bit patterns are assigned: one to indicate that quality of the synchronization is unknown and the other to signal that the section should not be used for synchronization. The remaining codes are reserved for quality levels defined by individual administrations. The remaining bits in Z1 and the Z2 byte are allocated for functions yet to be defined.

These definitions have introduced the concept of scrambling. Scrambling is used to ensure that the overall signal contains sufficient bit timing content. The scrambling algorithm is applied to the complete signal except for the first row of the STM-N SOH (9 × N bytes including the A1 and A2 framing bytes).

7.2.5 The pointer mechanism[9]

The mechanism that permits freedom of mapping of tributaries is the pointer mechanism associated with row 4 of the SOH, the AU pointer and the tributary unit pointers.

All SDH pointers have the same basic format. This is illustrated for the AU-4 pointer in Figure 7.9. In this case the pointer is contained in bytes H1, H2 and H3, row 4 of the section overhead. The information content of the

---

8. There seems to be a mistake here in G.708, which speaks of 82 bytes! This is presumably a mistyping of "B2".

9. Reference 113 from the index of references gives an alternative explanation.

pointer is contained in bytes H1 and H2, which are treated as a single 16-bit word. The first 4 bits of the word constitute a new data flag (NDF) and indicate the start of an entirely new data stream with the pattern "1001". Otherwise, the NDF is "0110". The next two bits indicate that this is an AU3, AU4 or TU3 pointer. The following 10 bits give the pointer value, a numerical value that indicates the offset to the start of the VC. This pointer value section has each alternate bit designated I (increment) or D (decrement).

The H3 bytes provide the opportunity for negative justification whereas the positive justification opportunity will be contained in the first three bytes of the AU-4. The pointer values indicate the start of a group of three bytes so that justification can occur no more frequently than once every three bytes.

Figure 7.9 introduces the concept of concatenation. Concatenation is used as an opportunity for doing the same with SDH as was described for wideband signalling in Chapter 5. When a signal is greater than the available bandwidth then several AUs can be concatenated together.

The working of the pointer mechanism depends on any pointer pattern

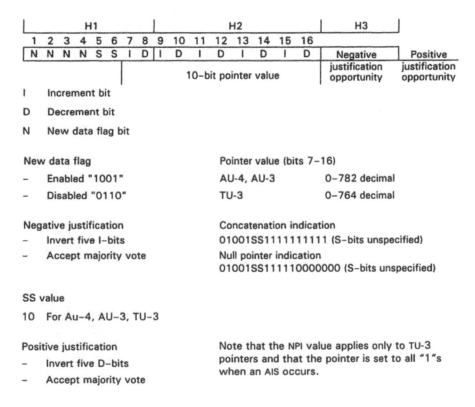

Figure 7.9 AU-n/TU-3 pointer (H1, H2, H3) coding (Figure 3-3 of Recommendation G.709).

remaining constant for three successive scans. It is illustrated for an AU-4 in Figure 7.10, positive justification, and Figure 7.11, negative justification.

Taking Figure 7.10 first, the NDF is "0110", indicating that this is not new data, and the pointer value is "*n*", pointing the offset to the start of the VC-4. At some point the frequency shift is such that the sending end recognizes that the offset should be moved; the same pointer value is sent but the I bits are inverted. The receiving end recognizes the inversion of the I bits as a signal to apply positive justification and inserts a positive justification group of 3 bytes as shown. On the next scan the sending end will increment the pointer value, thus pointing to the new offset directly. The new pointer value is "*n*+1". Should the frequency shift require negative justification, then a similar procedure occurs (Fig. 7.11), only this time the D bits are inverted, indicating to the

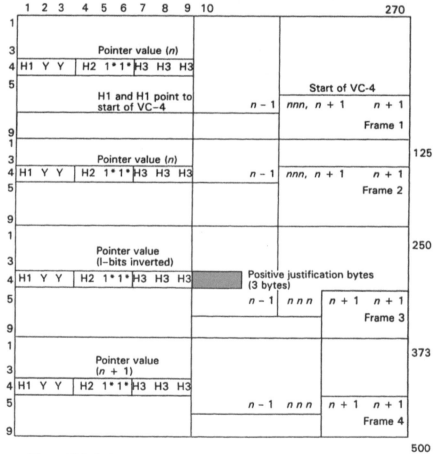

1*   All-1s byte

Y   1001SS11 (S bits unspecified)

Figure 7.10 AU-4 pointer adjustment operation, positive justification (Figure 3-4 of Recommendation G.709).

receiving end that a negative justification is required. In this case the H3 bytes are used for data, thus bringing the pointer offset forward by 3 bytes. On the subsequent scan the new pointer value, "$n-1$" is transmitted. Were either D bits or I bits to be inverted but accompanied by an NDF set to "1001" then the indication to change the offset would be ignored as this is the start of an entirely new envelope. The receiving end actually makes a decision if a majority of the five D or I bits are inverted.

This pointer mechanism works throughout the hierarchy using substantially the same format. The position of the pointer is different at different parts of the hierarchy. For this level of detail the reader is referred to the recommendations.

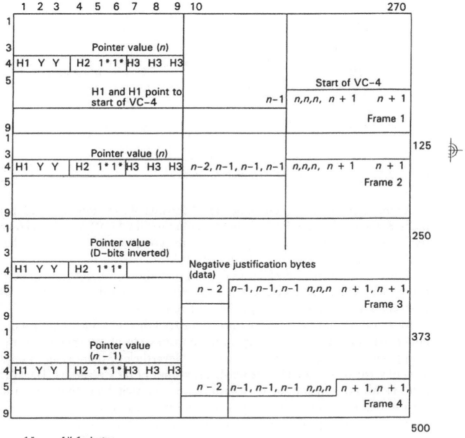

1*   All 1s byte
Y   1001SS11 (S bits unspecified)

Figure 7.11 AU-4 pointer adjustment operation, negative justification (Figure 3-6 of Recommendation G.709).

167

7.2.6 Summary

The explanations of the SDH multiplexing mechanisms have been based upon the diagrams included in the specifications. A further such diagram, taken from an early version of the SONET specification[116], is included as Figure 7.12 to serve as a basis for a summary description of the mechanisms.

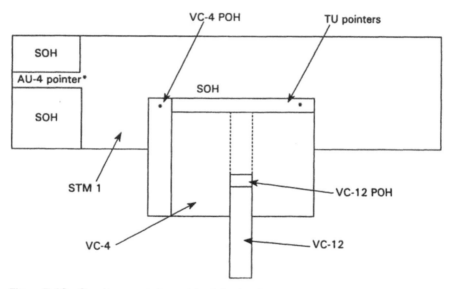

Figure 7.12 Synchronous transport module structure.

A synchronous transport module STM-1 consists of a section overhead that includes the AU pointers. Each virtual container within the AU contains its own path overhead and includes TU pointers to the component TUs. This mechanism can extend through more than one level.

There is therefore a mechanism defined that allows free manipulation of tributaries within the envelope but also permits a considerable amount of freedom in the choice of multiplexing structures.

Most important of all is the feature of the SDH that the content of the containers has absolutely no bearing on the transmission of information or the manipulation of the various data streams. Nevertheless, there is an in-built level of security provided by the parity provisions of the overheads. The SDH concept allows integration of transmission and switching functions and ensures that errors introduced by these transmission and switching functions are not allowed to proliferate into the data streams themselves.

Chapter summary

This has been a most enjoyable chapter to write. It is exciting to set out to describe, however inadequately, a technology that is very clever. There was much the same excitement in writing Chapter 3 of reference 2 describing the fundamentals of PCM. It seems probable that SDH is just as revolutionary as was PCM. Certainly, they both have this quality of engineering elegance and cleverness.

We started from a different point from the introductory material in Chapter 5, from the perception that the idea of pointing to the start of a message by using a digital coding of the position vector was already inherent in the PCM idea. It was also shown, using the UK as an example, that the availability of high bandwidths in optical transmission systems is causing administrations radically to change their plans for the network hierarchy.

It was then necessary to get down to the hard work of introducing the SDH in some detail, continually adverting to the wonderful pointer mechanism but covering a great deal of ground before finally getting to a description of it. This description has, I hope, shown its elegance. If the transmission is going rather slower that of late, stuff in an extra byte[10] while the mathematics of the transmission is updating the pointer. If, on the other hand, the transmission is going rather faster, steal a byte from the overhead to accommodate the signal while the updating is taking place.

---

10. It was an extra byte in SONET; in SDH of course, it is three bytes.

CHAPTER 8

Progress towards ISDN and B-ISDN

Progress (and its opposite) occur in areas other than the physical progress of technology. This account of progress will deal with three areas of movement, each of which has had, and continues to have, a profound effect on the development of the ISDN. The most obvious area in the first edition was the conduct of ISDN trials and this was treated first. In this second edition this treatment is curtailed and to it is added a short account of the present deployment of ISDN and the plans existing for the B-ISDN. Despite the lapse of time, not all of this account will be very encouraging.

Least obvious of all perhaps is the second area of communications industry reorganization. After very many years of comparative stability the industry has been undergoing, and continues to undergo, radical realignment in many countries.

The third area is that of standards, which is discussed in some detail throughout the book. In this chapter the developing mechanism for arriving at consensus international agreement on standards will be discussed.

8.1 ISDN and B-ISDN deployment

The concept of the ISDN became current thinking among telephone switching development teams in about 1980. The concept was received enthusiastically by manufacturers because it embodied attractive features assisting the marketing of switching products to administrations. The reasons for this attractiveness have been touched upon and will be dealt with again later. The supposed attractiveness of ISDN was (and is) particularly relevant to less developed networks in which the existing investment in plant and equipment is not endangered. The author remembers in particular the intense and very early interest in ISDN for applications in the proposed industrial and commercial developments in China.

In the developed networks, administrations were, with notable exceptions, slow to plan ISDN implementations, but anxious to participate in the definition deliberations. It is such enthusiasm for definition rather than trial implementation that resulted in the advanced, unproven state of the CCITT recommendations before significant experience of working trials.

It was therefore the manufacturers most committed to exporting to less developed networks and without extensive "home" markets that first engaged in concrete ISDN developments. Chief among these were Ericsson and ITT. It is only since mid-1984 that home-based manufacturers, AT&T, Northern Telecom and CIT Alcatel announced ISDN implementation plans. The first two of these were goaded into action partly by the enthusiasm of some of the US operating companies for ISDN trials. This enthusiasm was in turn partly fostered by foreign (to the US) suppliers whose home environment included a commitment of some sort to ISDN. Plessey (UK), Siemens (W Germany) and CIT Alcatel (France) were chief among these.

Excluded from this resumé of the early plans for ISDN, are the German manufacturers because the Deutsche Bundespost possessed clear but specialized ISDN plans involving an early move to the B-ISDN as it was then envisaged, the British manufacturers, again because the main network provider, BT, followed an individual route, and the Japanese for similar reasons.

Throughout the period of early trial implementations up to the date of the first edition of this book, there was a paradox that is still detectable today. Any cursory glance at the contents lists of the technical literature of the time indicates that the ISDN was receiving considerable attention, but more careful study of the material indicates much repetition of a relatively small amount of information, most of it drawn from the I-series recommendations. The only section of the industry that was consistently enthusiastic about ISDN was the private exchange suppliers, who actively marketed "ISDN" in the private network although compatibility problems between different products, because of the lack of agreed standards, discussed in Chapter 4, made "private" ISDNs problematical unless the same PABX system existed at all the major nodes of the private network.

A review of the account of these early trials and implementations indicates instances of enthusiastic activity followed by long pauses, and of trials completed but without published results or trials planned but with implementation delayed, all evidence of considerable shifts of opinion and strategy. The initial enthusiasm of the industry seems to have been short-lived and was followed by a very much more cautious approach. This variation in enthusiasm had regional modifications. The Germans moved from a plan for "broadband only" to broadband via "conventional" ISDN. Relative disinterest in the US changed to a wonderful variety of customer trials and, conversely, early firm intentions in the UK developed into a marked inactivity despite the existence of the only commercial ISDN service in the world. In defence of the UK attitude it must be remembered that the digital network was being rapidly implemented with full #7 signalling. All that was absent for ISDN implementation was suitable terminals.

171

The reason for this general lack of enthusiasm, in so far as they are known, are not all, or even mainly, technical. It is the object of the next part of this chapter to investigate the non-technical developments aiding, or hindering, ISDN progress. The technical difficulties are to be discussed in Chapter 9.

In Europe the creation of the European Telecommunications Standards Institute (ETSI) was inspired in part by the perceived need to have ISDN standards suitable for the whole community. Thus, the main DSS 1 standards are paralleled by ETSI documents[117, 118]. Formulation of these standards from ITU-T recommendations and provision of the accompanying testing and approvals documentation has been no small task, so that the Euro-ISDN has only become available in some countries in the European Union within the past three years.

Today ISDN service, to European standards equivalent to DSS 1, is available in the UK, France (Numeris) Germany and the Netherlands, but the marketing effort being expended in all but France is negligible.

Progress of the B-ISDN has not resembled the pattern of the ISDN. The contributing technologies are enjoying significant progress and SDH is being deployed in the trunk networks of the UK, USA and many other countries. SDH is also being used in the UK for some of the networks in the cable TV franchise areas in the UK. ATM is enjoying considerable attention as a high-capacity LAN in private networks and is being deployed over SDH in some trunk networks[119–122]. There is therefore a good chance that B-ISDN features can be deployed as soon as the terminal equipment is developed. The account of the ATM standards included in Chapter 6 indicates that most, but not all, of the conditions for deploying universally acceptable terminal equipment are now in place.

8.2 Network organization and the ISDN

It will be a question for technical historians to gauge the effect that changes in the organization of the telecommunications service have had upon technical development. That there have been massive changes is undoubted, and these changes have been coincident with the development of the ISDN. It is the intention of this section to investigate the effect of the one upon the other as a necessary part of the ISDN narrative. It is still too soon to hope that this account can adopt a historical perspective as the changes and the ISDN are still developing and still, it is suggested, interacting.

In 1980, when the ISDN idea first became popular, the USA was practically the only nation whose telecommunications were not controlled by an administration with close governmental links. While being unique in this respect, the American situation was still similar to that in other countries in being monolithic – AT&T controlled the large majority of manufacturing, development and operating companies. In the government-controlled telecommunications world it was received public and political doctrine that the telecommunications service was at best customer unfriendly, more usually atrocious. This was true even in countries such as France, whose network had leapt from being the worst in the developed world to the best within a decade. Telephones are like mothers-in-law – universally complained about with or without justification.

By 1980 a post-war mould had been broken by both inflation and interest rates. Rates to which the world had grown accustomed rose abruptly to double figures and generally remained at levels orders of magnitude in excess of the previous norm for the rest of the decade, and in much of the world until the present day. In these newly prevailing conditions, assets assume disproportionate importance and telecommunications networks are a very significant asset. It was, therefore, not surprising to find governments seeking ways to realize assets in order to finance the expenditures that caused the rise in inflation and interest rates in the first place and the telecommunications network became a prime target.

Thus there occurred the twin movements of liberalization and privatization: liberalization allowing new sources to compete with or enhance the service provided by the existing telecommunications asset and privatization realizing on the asset by converting its administration into a public company. The UK experienced both concurrently despite the inherent conflict between the two. Japan followed suit. The Europeans laboriously put together legislation that would ensure that the remaining government-controlled administrations would have no advantage in the open market and, when the Communist empire collapsed, the West insisted upon tying aid to a similar degree of liberalization and privatization that is proceeding with varying degrees of success.

In the US the same effect was achieved coincidentally with the UK by means of an entirely different mechanism. As a result of various anti-trust suits, the Bell monolith was compelled to divest itself into its component parts. As a result, the local network is now served by the Bell Operating Companies and the independents. The trunk network is available via several competing networks, of which AT&T still has the lion's share. By a separate mechanism – invasion, by Northern Telecom in particular, – the AT&T manufacturing arm is no longer as dominant.

In the US also, divestiture has been accompanied by liberalization because the role of the network provider has been closely defined, leaving room for service providers to offer their wares, as of right, over the network. It is therefore necessary to consider the effects on the ISDN concept of the emergence of: multiple networks; multiple services; and privately owned network providers. Rather than attempting to synthesize a global view of these effects, the discussion will be confined to a consideration of the history of UK privatization and liberalization in the hope that the lessons illustrated will have more general application.

8.2.1 Multiple networks

In 1980 the Post Office Corporation could view the advent of the ISDN in exactly the terms outlined in Chapter 1: as a means of deriving more revenue from the existing network by enhancing the services available over that network. Savings could also be made by absorbing other networks (e.g. telex) within the ISDN.

Liberalization brought to the UK a rival network [Mercury Communications Ltd (MCL)], freedom to the user to purchase (approved) PABXs from any source, and a compulsion on BT to allow value-added services to be offered over its network components. The present position is that there are some six network providers offering rival trunk services, these or others offering competing international switched or leased line services and the cable TV companies offering a rival local service within their franchise areas.

Privatization brought to the UK a dominant network provider (BT) half-owned by its shareholders and compelled to compete even-handedly with its competitors limited, for five years, to Mercury only.[1]

Privatization lost to the UK a competent, authoritative voice in international forums. The seats at CCITT, CEPT and other meetings previously occupied competently by BT had to be filled by government servants or their nominees and some, at least, of the competence and authority were inevitably lost. These new incumbents, even when they were BT personnel nominated by government, were hampered not just, perhaps, by lack of experience, but by the need to speak for the divergent interests of competing network providers. It is the author's impression that the UK contribution to international debate has, as a result, become less authoritative and considerably more shrill with self-interest. In the fullness of time it was presumed that this vacuum would be filled by the Office of Telecommunications (OFTEL), set up to police privatization, but this has not proved to be the case. However the representation is arranged, there will always be an innate disadvantage for the representative acting for several network operators compared with one who represents government and administration at one and the same time.

Strategy is determined by a consideration of the best interests of the organization. The best interests of a sole network provider are very different from those of a network provider in competition. In the latter case interests and therefore strategy will be determined in part by the strategy of the opponent(s). Mercury, quite rightly, went first for the rich pickings from major business in the main conurbations by offering high-capacity digital network connectivity over a few highly secure, very popular routes.[2] BT, in response, offered Kilostream (64 kbit/s) and Megastream (2.048 Mbit/s) digital private

---

1. And, of course, Kingston-upon-Hull Corporation, now Kingston Communications, which had been an independent network since 1910.

2. Mercury provided rapid local network service to the City of London by obtaining existing ducts for their cables by astutely purchasing the redundant pneumatic

circuits. Interest, within BT, in the ability to offer 144 kbit/s access to individual subscribers and partial access to 2.048 Mbit/s primaries to small business decreased, particularly because of the absence of suitable terminals or PABX systems.

Privatization also brought to the UK the need to police the competing networks and liberalization brought the need to gain approval for network attachments and network-provided services. Approval had always been a requirement, but one that devolved as a duty upon the sole network provider who could conduct the approval exercise against standards produced by itself and in accord with international standards where this was appropriate. The need for independent standards meant a massive increase in the national standards bodies. Most of the responsibility for the organization of new standards work was given to the British Standards Institution (BSI), but the members of the various committees were, largely, those same people from BT and the manufacturers that might otherwise (more usefully?) be serving on the international bodies, leading to a further attenuation of the UK voice in international discussions.

The moves by the European Union (EU) towards open network provision (ONP) have resulted in much, but by no means all, of this national standards activity being transferred to Brussels, or Sophia Antipolis, where the European Telecommunications Standards Institute (ETSI) is based.

8.2.2 Multiple services

In Chapter 1 the concept of varied services provided over the telephone network was introduced. One of the first examples in the UK was the commercially premature introduction of Prestel by BT in the UK. By the early 1980s the concept had gained popularity and there are now many such services available. At the time of privatization the concept was formalized and just as BT, Mercury and Hull were licensed to provide public network services so too licences were issued and continue to proliferate for value-added network services (VANS) or value-added data services (VADS). A VANS is much broader in concept than a Prestel-like service and was initially defined to embrace any service that "added value" to a communication over and above the transport function provided by the network.

Thus, a service switched in to a communication to provide a higher level of data integrity, to encrypt, to translate between protocols or to store and forward messages, would be a VANS. A service to provide calling line identity plus, perhaps, name, address, credit card number, etc. to the called party desiring it (a mail order firm?) could be a VANS.

---

ticket tube system that used to interconnect all the main manual switchboards and all the major business premises in the city. The author's father had been engaged in the installation of this system when he first arrived in London in the early 1920s.

175

Table 8.1 Examples of VADS/VANS applications.

Electronic data interchange (EDI)	Trading interchange for procedures such as: Order intake Order acknowledgement Invoicing
Electronic mail systems (EMS)	Person-to-person messaging using mailbox techniques
Databanks	Access to business information. Continuous information update in databank and, if desired, to user.
Electronic fund transfer	Automated transaction systems.

Table 8.2 Examples of VADS/VANS service providers.

Service	Supplier	Applications	Network
Prestel	BT	Databank	Subscriber access via PSTN
Telecom Gold	BT	EMS	Subscriber access via PSTN
Fastrak	Travinet (of Midland Bank)	EDI/EMS/Databank	Data line to own network
Tradanet	ICL (and Article Numbering Association)	EDI	Data line to own network
Infotrac	Istel	EDI/EMS/Databank	Data line to own network

By 1987 there were more than 600 licensed VANS and VADS serving the UK. Prestel (videotex) and Telecom Gold (teletex, perhaps, or just E-mail) are examples of BT offerings in the field, and BT has many competitors offering similar services. Table 8.1 gives details of a few VANS and VADS applications and Table 8.2 details a few of the commercial offerings in the UK. Today the figure is about the same, some of the early entrants having proved unsuccessful. It is arguable that earlier implementation of the ISDN would have assisted the progress of VANS.

The penny will, of course, have dropped; the perceptive reader has already noted that the very services that were anticipated as advantages of the ISDN have emerged on the present network in advance of the implementation of the ISDN. The ISDN can of course greatly simplify the access to these services, but the advantage that would previously have been shared between the user and the sole network provider is now spread over a multiplicity of claimants. The attractiveness of the ISDN concept to the network provider who must implement it has been considerably diluted.

A new service that demands mention here, although its description as a VANS is controversial, is the cellular radio service. The Scandinavian countries, with the Nordic network, were the first to enhance mobile radio from a manual or semiautomatic system to a completely automatic system, allowing calls to be made and continued to and from a mobile user regardless of the movement of the mobile. The UK cellular services were initiated in January

176

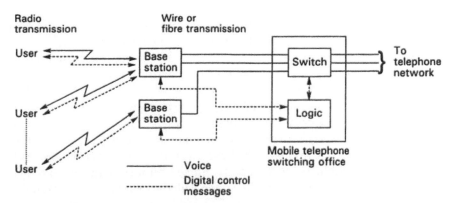

Figure 8.1 Principle of the cellular telephone[80].

1985 by two rival consortia, BT/Securicor and Racal Vodaphone, and immediately exceeded expectations with 60000 and 65000 users respectively by December 1986. The cellular networks are separate digital networks using digital exchanges at the switching nodes but with access to and from the PSTN. The cellular exchanges are equipped with an additional facility, called "roaming", to track the mobile and "hand on" the communication link by transferring the connection from the transmitter of the cell just being left to the transmitter of the cell about to be entered (Fig. 8.1).

This fresh population of, presumably, sophisticated users is a prime candidate for the ISDN. The original cellular networks were implemented using known radio techniques: frequency division multiplexing and frequency modulation. The more recent offerings, now implemented as parallel cellular networks, are digital. Clearly, therefore, only the upgraded networks will accommodate the cellular ISDN.

8.2.3 Privately owned network providers

At its inception Mercury was owned by a partnership of Cable and Wireless, British Petroleum, Barclays Bank and British Rail, the last of which provided the wayleave for the networks' fibre trunks laid alongside its track. Cable and Wireless later bought out the other shareholders and now has an 80% share in MCL.

Privatization of BT left it still 51% government owned and with the largest public shareholding by small investors of any UK company.

The Hull network is wholly owned by the Kingston-upon-Hull Corporation and has always been both a thorn in BT's side (and of the Post Office before it) and a source of fresh impetus to both technology and services.[3] In

---

3. Hull was the first to offer dial-up services such as dial-a-disk, shopping information, etc.

recent years the telephone department has been separated from the local authority to give the former greater commercial independence and flexibility.

After BT was created as a separate corporation from the Post Office in 1982 and until privatization in 1985 the annual publication of BT accounts was a source of considerable embarrassment because of the "huge" profits. These were not in fact excessive when related to the assets and investment used, but nevertheless were the occasion for considerable public outcry. Since privatization the annual reports have shown much improved (greater) profits and little adverse comment has resulted. The business attitude of BT has altered just as radically as the public perception of BT. All new business ventures, and there are many, are assessed and embarked upon with the clear intention of improving BT profitability as a first priority. The service imperative is certainly still present and important but has ceded to profit as the primary objective.

To envisage a universal ISDN therefore, BT has to consider first profitability, including possible competitive edge over its rivals. As the dominant network provider it has to consider too its statutory obligation to provide access between BT users and users of competing networks. Granted this access and assuming that users on other networks are also provided with the competing network's version of ISDN, is it in the overall interests of BT to provide enhanced ISDN features to such users? The solution to these problems is reasonably simple, and probably favours BT and the ISDN, but there is a clutch of related problems concerning charge apportionment (to be dealt with in Chapter 9) that are complicated by the existence of (at least) two integrated networks or two portions of the ISDN.

As a private company, BT could no longer depend on the UK operation for continued growth in profits. This is particularly so because the regulator, OFTEL, immediately imposed a ceiling on BT prices based on a formula of RPI $- x$%. A price cap such as this is still in force today and the proposal for the period of five years from 1998 is at present being negotiated. Furthermore, OFTEL insists that all the network providers, particularly BT, must not subsidize loss making operations from profit from other areas of activity. For growth, BT had to look elsewhere and did so with international alliances with other carriers. Just as this book is going to press, BT and Cable and Wireless have resumed discussions of a merger that have been proceeding intermittently for the past several years. Such a merger would provide BT with the diversity of Cable and Wireless's interests in world-wide, notably Asian, networks. Such a merger would also involve the sale by the new company of MCL as the UK government and regulator would not tolerate the departure from liberalization involved in BT again becoming almost the sole network provider. Among the concerns thought to be likely to bid for MCL is AT&T, which is now, as a result of American liberalization, able to seek replacement for the international business denied it in 1925.

It would be a Herculean task to synthesize from the UK experience a world view of organizational changes in telecommunications. It suffices to observe that the UK is not alone, although in the lead, in developing along the lines described. If there are lessons to be learnt by other national networks following similar routes they should be learned now. What would be useful to examine, however, is the equivalent situation in the most developed and the biggest national network in the world.

8.2.4 The American experience

It is remarkable that the US network, starting from a radically different situation, has developed in such a way that it inherits many of the problems discussed in relation to developments in the UK.

While always being a privately owned network, the US was so dominated by the AT&T organization that AT&T received the status and attention due to a national network administration. This being said, there were always separations in intention if not in fact. To maintain the status of the independent operating companies (providing in 1980 some 10% of the local network) the independents and the Bell Operating Companies (BOCs), then a part of AT&T, were unable to provide trunk network service but could obtain this from independent trunk operators, of which AT&T Long Lines was, and is still, the biggest. As described in reference 2, the (AT&T) trunk network has adopted CCS almost universally but, in the early stages, this was mostly the analogue version, #6 signalling.

Divestiture of AT&T resulted in the BOCs becoming private, independent entities, in the formation of a national research institute serving the operating companies and equivalent to Bell Laboratories (AT&T), Bell Communications Research (Bellcore), and in the firm determination to allow complete user freedom to connect any (approved) device to the network (see Computer II, Chapter 3).

AT&T was actually broken up into seven regional entities each consisting of several operating companies to a total of 22. Each company is the sole provider of telephone services within a local access and transport area (LATA). The full list, which totals 19, the remaining three organizations being service organizations, reading from west to east across the continent, is as follows: Pacific Telesis Group (Nevada Bell, Pacific Bell); US West (Mountain Bell, Northwestern Bell, Pacific Northwest Bell); Southwestern Bell Corporation (Southwestern Bell); Ameritech (Illinois Bell, Indiana Bell, Michigan Bell, Ohio Bell, Wisconsin Bell); Bellsouth (South Central Bell, Southern Bell); Bell Atlantic (Bell of Pennsylvania, C&P Tel, Diamond State Tel, New Jersey Bell); Nynex (New England Bell, New York Tel). In April 1996 Bell Atlantic and Nynex announced their intention to combine provided that the US regulatory authorities permit this.

The American user, unlike a UK user, therefore has no choice of local network as the operating company enjoys a local monopoly, but has a choice of

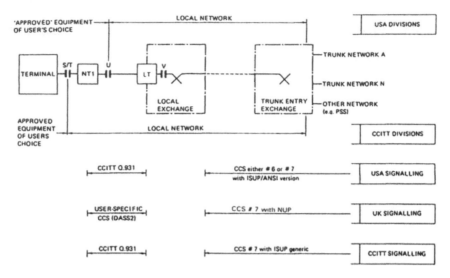

Figure 8.2 Network arrangements for ISDN.

trunk network [unless the operating company can convince the Federal Communications Commission (FCC) that it is technically incapable of providing a choice of carrier] and has a choice of device to connect to the network. As in the UK, this freedom to choose terminal devices has directly encouraged the creation of VANS that in America have achieved a greater degree of penetration than anywhere else in the world.

Figure 8.2 shows the effects of these organizational arrangements upon the USA network in contrast to what is becoming the accepted norm in the rest of the world.

The American upheaval is still continuing. Because the regulations concerning the use of private networks were less strictly formulated than in the UK, it is possible for American users to access the trunk network via their private network, thus bypassing the local operating companies. There is even a trade in selling on "bypass" facilities to other users. Largely because of the presence of bypass, the operating companies have secured relaxation of the Computer II distinction between enhanced and basic service and Computer III (June 1986) substituted "competitive" and "non-competitive" for the words "enhanced" and "basic". The operating companies are now free, for example, to offer X.25 protocol conversion as a "non-competitive" service.

The USA now enjoys telecommunications equipment and services provided by some 8000 equipment suppliers and some 200 network service vendors. Liberalization indeed!

8.2.5 Changes in manufacturing

Massive changes have also occurred in the organization of the manufacturers. These changes are also germane to our subject and are outlined in a companion volume[2].

8.3 Setting the standards

International telecommunications, practically alone in international relations activities, has been conducted successfully and amicably since 1868 on a basis of unanimous consensus agreement. The International Telegraph Committee (CCIT), originated in that year, grew with the technology it controlled into the CCITT (telephones and telegraphs) and the CCIR (radio), and these in turn became part of the ITU when the communicators were assumed as agencies of the newly formed United Nations. At a plenipotentiary conference in Geneva in December 1992 it was agreed to restructure the ITU organization. The most immediately significant result of this has been that CCITT and CCIR recommendations have become ITU-T and ITU-R recommendations respectively.

It had been the practice of the ITU for very many years to arrange its affairs into a four-year cycle. During each period, outstanding questions and new problems were considered by study groups and each cycle culminated in a plenary session at which the new and revised recommendations agreed during the period were unanimously ratified. The recommendations current at the time of the first edition, the Red Books, were the product of the eighth such plenary assembly at Malaga–Torremolinos in 1984. These were superseded by the Blue Books as a result of the ninth plenary in Melbourne in 1988. The Blue Books will perhaps be the last occasion when all the ITU-T recommendations are published at one time in a single set of volumes. Subsequent decisions, at the 1992 plenipotentiary and previously at the World Administrative Telegraph and Telephone Conference (WATTC), also held in Melbourne in 1988, changed the pattern to that used by most other standards bodies of issuing new and revised documents as soon as they have achieved international assent.

The subject matter of CCITT recommendations was confined, until quite recently, to areas pertinent to international communications. Thus, CCITT could define a budget of loss for an international telephone connection but could make no recommendation on how the portion of the budget destined for the national part of the connection could be apportioned. Similarly, CCITT R2 Signalling is defined for international use only, leaving considerable freedom for national variation. This deliberately confined stance of the CCITT, often criticized by engineers finding difficulty reconciling their systems to national variations, has, in the case of the ISDN, been abandoned in favour of attempts to establish uniformity throughout the network and not only, or even principally, across frontiers.

It has already been noted that this penetration of the CCITT remit has first been evident in digital network definition (G.732 and others) followed coincidentally by CCS #7 (Q.701 and others) and by data communication (the V series and X series) and now by the I series on the ISDN. A parallel and no less significant change from specification production after the event to specification before development has also been noted already. This second change is important enough to warrant an illustration.

Reference 2 (among many) describes how PCM development followed different routes in North America to the route subsequently adopted by the rest of the world with the result that CCITT had to bow to the inevitable and define two standards post hoc. In contrast, the ISDN access recommendations have been defined in some detail before any but very tentative trial implementations. It can be argued, and the contention will be supported by argument in later chapters, that this ante hoc approach results in less elegant, more cumbersome solutions, places a dead hand on imagination and invention, and delays development. In theory at least, however, ante hoc specification does remove the dangers of finishing up with two or more, inconsistent systems to specify for worldwide application.

The extension of the ITU remit has been caused, in part, by popular demand. A significant aspect of this demand has been the fragmentation, already noted, of the network provider authorities. This same fragmentation has necessitated an elaboration of national "pre-ITU-T" forums, part of whose functions are to brief the real ITU-T representatives.

In this section, the organization of the ITU and the other international bodies that contribute to its work will be described briefly and the "pre-ITU-T" national bodies in both the USA and the UK, the nations where fragmentation has proceeded furthest, will also be considered, again briefly.

8.3.1 International bodies

Figure 8.3 provides an overview of the international bodies chiefly concerned with communications. This overall view is elaborated for the ITU in Figure 8.4, for CEPT in Figure 8.5 and for ISO (of which only one technical committee, TC 97, the author of ISO 7498 on OSI, need concern us) in Figure 8.6. The ECMA organization, again involving just one technical committee, TC 32, of interest here, does not warrant a separate diagram. ECMA, nevertheless, is playing a very important role in compelling acceptance of OSI by the other international bodies and by the less enthusiastic (for standardization) computer suppliers. ECMA is perhaps the catalyst in the international standards "reaction".

The detail of Figure 8.4[123] reveals the ITU study groups most concerned with ISDN and related matters. These are indicated in the diagram. The

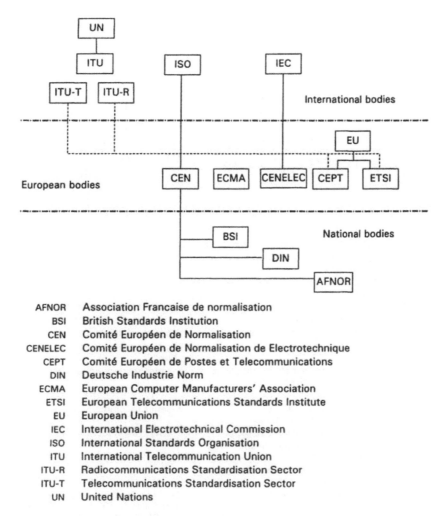

AFNOR   Association Francaise de normalisation
BSI   British Standards Institution
CEN   Comité Européen de Normalisation
CENELEC   Comité Européen de Normalisation de Electrotechnique
CEPT   Comité Européen de Postes et Telecommunications
DIN   Deutsche Industrie Norm
ECMA   European Computer Manufacturers' Association
ETSI   European Telecommunications Standards Institute
EU   European Union
IEC   International Electrotechnical Commission
ISO   International Standards Organisation
ITU   International Telecommunication Union
ITU-R   Radiocommunications Standardisation Sector
ITU-T   Telecommunications Standardisation Sector
UN   United Nations

Figure 8.3 International telecommunications bodies.

urgency of the work on ISDN and related areas, and its quantity, led the study groups concerned to envisage the issue of interim Grey Book recommendations halfway through the 1985–88 study period. Progress was not sufficient to achieve this, but the idea took root and has led to the present arrangements for issue of recommendations as they are agreed.

It has been shown repeatedly how the international bodies have been interrelating on ISDN matters. CCITT republication of the OSI standard ISO DIS 7498 as Recommendation X.200 *et seq.* is but one example. All the bodies shown in Figure 8.3 now have fairly formal relationships as well as detailed informal co-operation in order to ensure that their pronouncements are harmonious.

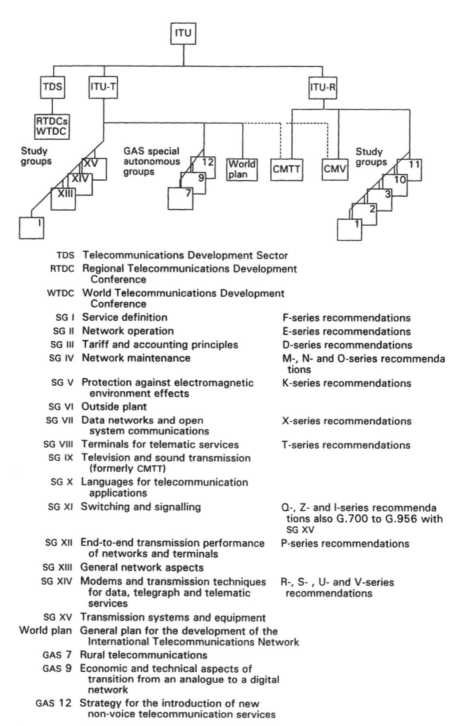

TDS	Telecommunications Development Sector	
RTDC	Regional Telecommunications Development Conference	
WTDC	World Telecommunications Development Conference	
SG I	Service definition	F-series recommendations
SG II	Network operation	E-series recommendations
SG III	Tariff and accounting principles	D-series recommendations
SG IV	Network maintenance	M-, N- and O-series recommendations
SG V	Protection against electromagnetic environment effects	K-series recommendations
SG VI	Outside plant	
SG VII	Data networks and open system communications	X-series recommendations
SG VIII	Terminals for telematic services	T-series recommendations
SG IX	Television and sound transmission (formerly CMTT)	
SG X	Languages for telecommunication applications	
SG XI	Switching and signalling	Q-, Z- and I-series recommendations also G.700 to G.956 with SG XV
SG XII	End-to-end transmission performance of networks and terminals	P-series recommendations
SG XIII	General network aspects	
SG XIV	Modems and transmission techniques for data, telegraph and telematic services	R-, S-, U- and V-series recommendations
SG XV	Transmission systems and equipment	
World plan	General plan for the development of the International Telecommunications Network	
GAS 7	Rural telecommunications	
GAS 9	Economic and technical aspects of transition from an analogue to a digital network	
GAS 12	Strategy for the introduction of new non-voice telecommunication services	

Figure 8.4 ITU organization.

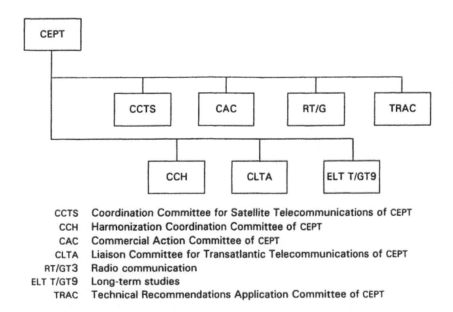

CCTS   Coordination Committee for Satellite Telecommunications of CEPT
CCH   Harmonization Coordination Committee of CEPT
CAC   Commercial Action Committee of CEPT
CLTA   Liaison Committee for Transatlantic Telecommunications of CEPT
RT/GT3 Radio communication
ELT T/GT9 Long-term studies
TRAC   Technical Recommendations Application Committee of CEPT

Figure 8.5 European Conference of Postal and Telecommunications Administrations.

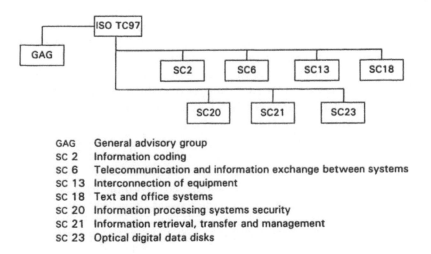

GAG   General advisory group
SC 2   Information coding
SC 6   Telecommunication and information exchange between systems
SC 13   Interconnection of equipment
SC 18   Text and office systems
SC 20   Information processing systems security
SC 21   Information retrieval, transfer and management
SC 23   Optical digital data disks

Figure 8.6 International Organization for Standardization.

It is worth emphasizing that, of all the international, regional and national bodies concerned with telecommunications and ISDN-related topics, only the ITU uses a system of consensus agreement. All the other bodies use some kind of voting technique. In practice this means that the ITU, even under the old dispensation, tends to produce standards rather faster than the other bodies.

185

Compare the previous four-year plenary cycle of the ITU with, say, the ISO performance of 6–7 years. The other side of this coin is that the ITU will happily issue a recommendation with gaps, as we have seen, "for further study". An ISO or BSI document, by contrast, will, eventually, appear whole and entire. The difference is that the latter are standards and their provisions are considered as mandatory whereas the ITU documents are recommendations and, although their provisions have considerable force and will influence the content of many of the other standards, they are not mandatory.

8.3.2 European bodies

In the first edition the equivalent to Figure 8.6 did not include the European Telecommunication Standards Institute (ETSI), which at the time of writing had not been created. Now it is an important additional resource in the international process of arriving at standards in telecommunications.

One of the driving forces towards liberalization has been the European Economic Community (EEC) now known as the European Union (EU). Its requirements for open provision of goods and services throughout the Community met with resistance in the telecommunications sphere because in each country a terminal had to satisfy a different set of technical standards before it was allowed to be connected to the network. This meant that the local industry had a distinct advantage and telecommunications exports between member countries had been low to vanishing point as a result.

In 1988 the EEC supervised the creation of ETSI charged to produce European Norms to set common requirements for the whole of the Community. Some of these standards may then be used, where necessary, to form the basis of legally enforceable regulations.

ETSI operates in much the same way as the ITU through a number of committees and subcommittees analogous to the CCITT study groups. Like most other standard- making bodies other than the ITU, ETSI relies on a voting procedure that, like most things in Europe, is weighted according to the economic "weight" of the member country. At earlier stages, however, consensus is required so that a draft standard must be agreed unanimously by its parent committee before being published for public comment and adopted by weighted vote of the members.

8.3.3 National bodies

It is sufficient, and probably only possible, at this stage of the development of national/international relationships, to examine the arrangements in the UK and the USA to harmonize submissions to the international levels. As the fragmentation caused by deregulation, privatization, liberalization, etc. progresses, no doubt the picture will become more complex and similar international arrangements will emerge in other countries.

8.3.3.1 The USA

The organizations contributing to telecommunications standards-making in the US are shown in Figure 8.7, which depicts the US CCITT National Committee in some detail. Figure 8.8 performs the same function for the Exchange Carriers Standards Association (ECSA) and Figure 8.9 shows detail on the other contributing bodies. Some of these bodies have been created as a result of the perceived importance of the ISDN. Some have appeared as a result of deregulation; the national committees and ECSA, which nevertheless pre-existed deregulation, are examples.

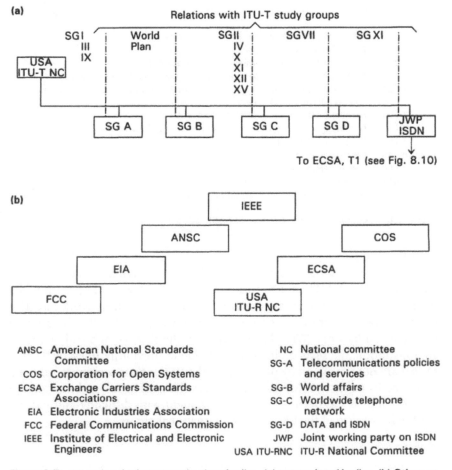

Figure 8.7 US national telecommunications bodies. (a) ITU-T-related bodies. (b) Other national bodies.

187

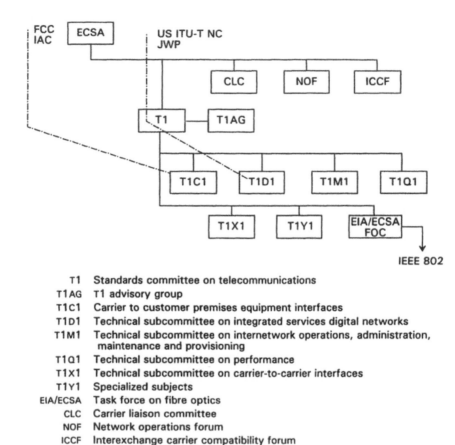

T1	Standards committee on telecommunications
T1AG	T1 advisory group
T1C1	Carrier to customer premises equipment interfaces
T1D1	Technical subcommittee on integrated services digital networks
T1M1	Technical subcommittee on internetwork operations, administration, maintenance and provisioning
T1Q1	Technical subcommittee on performance
T1X1	Technical subcommittee on carrier-to-carrier interfaces
T1Y1	Specialized subjects
EIA/ECSA	Task force on fibre optics
CLC	Carrier liaison committee
NOF	Network operations forum
ICCF	Interexchange carrier compatibility forum

Figure 8.8 Exchange Carriers Standards Association.

8.3.3.2 The UK

Figure 8.10 shows the situation in the UK in rather less detail than that adopted for the USA. Upon privatization of BT, OFTEL was set up to shoulder the responsibilities, previously assumed by BT, for overall network integrity. International co-operation, on the other hand, was passed to the Department of Trade and Industry (DTI). Nevertheless, in the absence of a truly national administration, the seats on international bodies previously filled by BT personnel are now occupied by people from the DTI or their nominees, advised, no doubt, by OFTEL, BT, Mercury and others. This ought only to be an interim solution and, as has been discussed, it is fraught with difficulty and has undoubtedly weakened the UK voice in international affairs.

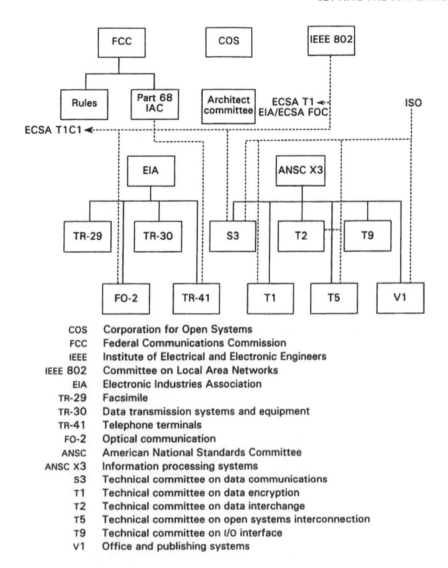

COS   Corporation for Open Systems
FCC   Federal Communications Commission
IEEE   Institute of Electrical and Electronic Engineers
IEEE 802   Committee on Local Area Networks
EIA   Electronic Industries Association
TR-29   Facsimile
TR-30   Data transmission systems and equipment
TR-41   Telephone terminals
FO-2   Optical communication
ANSC   American National Standards Committee
ANSC X3   Information processing systems
S3   Technical committee on data communications
T1   Technical committee on data encryption
T2   Technical committee on data interchange
T5   Technical committee on open systems interconnection
T9   Technical committee on I/O interface
V1   Office and publishing systems

Figure 8.9   Other US national bodies.

8.3.4 Review

The discussion has described, and justified, the need for standards-making bodies to become far more involved in technology because technology's advanced complexity requires increased control. It has justified also the need for standards to increase the areas of free markets in the provision of telecommunications networks, services and terminals. However, not all the results of this increased involvement have been happy.

189

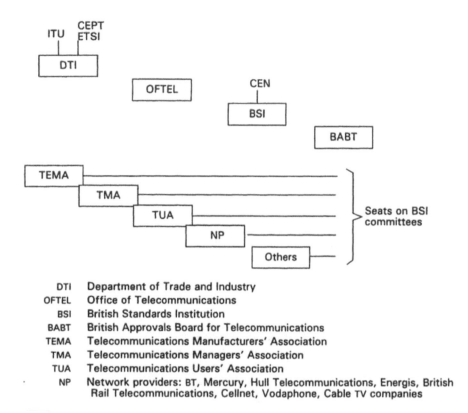

DTI   Department of Trade and Industry
OFTEL Office of Telecommunications
BSI   British Standards Institution
BABT   British Approvals Board for Telecommunications
TEMA Telecommunications Manufacturers' Association
TMA   Telecommunications Managers' Association
TUA   Telecommunications Users' Association
NP   Network providers: BT, Mercury, Hull Telecommunications, Energis, British Rail Telecommunications, Cellnet, Vodaphone, Cable TV companies

Figure 8.10 UK national bodies concerned with telecommunications.

Because the network is to be universal and the complex terminal is to be connected to the network at any point in the world, the terminal specification and network protocol must be compatible so that terminals may interoperate across the worldwide network. This is not a simple task to ensure, and the present attempts at solution are biased towards design of the protocol by international committee *before* any trials of tentative realizations of possible protocols. The international bodies, hugely successful in standardizing telecommunications up to the present, are being asked to perform design tasks more suited to dedicated design teams uncluttered by considerations of national or organizational interest.

ETSI has perhaps provided the germs of a solution to this problem. In areas where a distinct body of work is known to be required in order to produce a standard, ETSI recruits and hosts a team of experts, a project team, which is supervised by the parent committee and whose output will be considered for approval by that committee. This could well be a suitable pattern for future development projects when individual organizations are too unsure of the direction standardization will take to make significant investment on their

190

own account. At present, such ETSI project teams are confined to development of protocols for standardization, equivalent to paper design, although in areas such as conformance testing, project teams (not all of them ETSI) have developed the complete testing software. There is no good reason why, should the ETSI members agree, ETSI project teams could not be used for conceptual development projects funded by the members.

Chapter summary

Progress was discussed in the most obvious areas of actual ISDN implementation. Nowhere had this progressed beyond the trial stage up to 1988 except for the UK, which, uniquely, provided a tariffed service. The account illustrated that initial enthusiasm among the exporting manufacturers flagged somewhat, to be replaced by enthusiasm shown mainly by the American operating companies and by France Telecom.

One reason for this change in the identity of ISDN enthusiasts has undoubtedly been the coincidental changes in the organization of telecommunications in the US, Japan and Great Britain and, to a lesser extent, worldwide. These changes and advances in the technology of data communications have led to the introduction of ISDN features in advance of the ISDN and the consequent removal of at least part of the ISDN *raison d'être*.

These organizational changes have, in turn, had a profound effect on the processes involved in reaching consensus on standards. The final part of the chapter resorted to pictorial narrative to give some small idea of the resulting complexity of the standards-making process.

CHAPTER 9

The technical risks of the ISDN and B-ISDN

9.1 Introduction

Having negotiated a long section of narrative with much political content it is now time to approach the ISDN from the point of view of the designer and network planner seeking to implement a complex, new concept that is, as has been seen, at once insufficiently defined for planning and excessively specified for imaginative design.

Underlying the discussion until now, and central to the subject areas yet to be covered, is the fundamental dilemma first introduced in Chapter 3. We are seeking to replace universal network access to a reasonably simple device, the telephone (Fig. 3.5), with access to highly complex devices (Fig. 3.4). Indeed, the ITU-T access recommendations imply simultaneous access by up to eight such devices (Fig. 3.3).

In devising solutions to this dilemma the universal requirement leads us into further problems because of the two world digital standards, 24-channel G.733[114] in North America and South-East Asia and 30-channel G.732[19] in the rest of the world. It is not surprising that the difference in regimes that caused the G.732/733 dichotomy introduces additional problems in the path of universal ISDN.

This chapter attempts to outline most of the technical issues; the majority of them are indeed problems, but a few, notably Centrex, are far from being problems and may well prove to be technical catalysts of the ISDN.

Practically all the issues to be addressed are, in part, affected by continuing international and intranational discussion.

9.2 The U interface

The U interface was discussed in Chapter 3, in which it was revealed how the FCC insistence that customer premises equipment (CPE) should be freely available from any source introduced an interface (at the incoming PSTN pair of wires) that was not envisaged by the majority at the CCITT. It was hoped that Notice of Proposed Rulemaking, Computer III (August 1985) might encourage a change of heart on the part of the FCC. This was in spite of the fact that the Notice expressly discounts any intention to change the status of CPE. The Computer III report (June 1986) does indeed reveal that many comments supported such a change but the FCC has stuck to its guns and maintained the distinction. The report does, however, make allowance for multiplexer equipment belonging to the network provider being installed on the customer's premises in order to deliver a number of U interfaces to the customer.

Despite this minor retreat, it is now up to the ITU-T to attempt to move towards the FCC position. It is possible that this movement will take place (although no significant movement has occurred during the past eight years since writing the first edition) and the U interface will at last receive formal international recognition. Regardless of the FCC position, it seems evident to the author that refusal to define such an interface is to maintain an unnatural distinction against all logic.

A curious, analogous, incident took place in the UK. For IDA implementation, BT defined and commissioned several network terminals. One of these, NTE 1, was conceived and successfully developed as a completely self-contained digital telephone. As such it contained all the functions illustrated in Figure 3.6 (relevant to speech telephony) within the one piece of apparatus. OFTEL, on being offered NTE 1 for approval, rejected it on the grounds that the S/T interface was not available for the customer to plug in competitive equipment. This appears to be the same argument as that used by the FCC in support of a different interface. The result has been that a simple digital telephone for use on the IDA was not made available.

This is a dilemma that is part political, part technical. Without the U interface all the "network" features of NT1 – test, loop-back, etc. – are the responsibility of the network provider and can be locally specified by the network provider to suppliers. With the U interface, every supplier of CPE must be constrained to provide a basic set of universally identical network features. This in turn leads to the ability to design LSI NT1 equipments for much higher volumes than would be possible without U interface definition.

9.3 Error control[1]

Chapter 4 showed how the tendency to discard #7 as a data transport medium in favour of, for example, X.25-based data transport has led to the acceptance of a rather rudimentary error control mechanism. There are very

---

1. There is some confusion possible in discussing error mechanisms. For this discussion error control is taken to mean any function for ensuring delivery of error-free messages. The two main error control methods in use are error recovery (detection of

many error control mechanisms in existence. Many of the protocols mentioned in passing in previous chapters, CO3, SNA, X.25, etc., contain provision for recovery from errors between communicating machines. The majority of those that have been dealt with here are based on the use of HDLC. Error control is a two-part function: error detection and error correction by either retransmission or direct correction. Failure to achieve an acceptable level of error correction should eventually result in connection failure and consequent alarm. A lower level of alarm should, of course, also be generated by the existence of an excessive error rate even although it is still within the limits with which error correction can cope.

Suppose, for the sake of argument, that administrations were using their #7 signalling network for the transparent transport of data. Suppose too that the data had already been "packaged" in a protocol such as X.25. A possible connection path that would result in this eventuality is illustrated in Figure 9.1a. Now consider the resulting complete message and its treatment illustrated in Figure 9.1b. The data message has, at the PAD providing entrance to the PSS network, or at some earlier point, been packetized as illustrated in Figure 2.18. Indeed, there has been a LAPB (or LAPD?) "conversation" such as that illustrated in Figure 2.17 (or, in summary, in Fig. 9.2) preceding the transfer of real information. Each packet is accompanied by the two-octet check pattern obtained from the packet contents in exactly the same way as #7 constructs

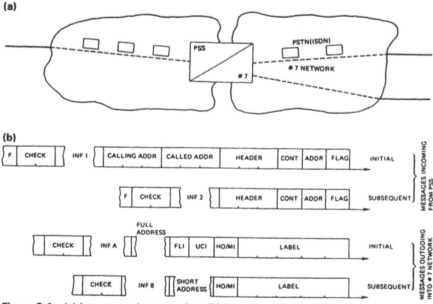

Figure 9.1 (a) Internetwork connection. (b) Internetwork messages.

error followed by recovery by means of retransmission) and error correction (detection followed by correction performed at the receiver by means of internal evidence in the, highly redundant, transmitted message).

194

the check bit from the #7 message (both protocols use HDLC). Unlike #7, the error recovery function is less sophisticated, the particular frame at fault is not identified, and a deficient count has to be responded to by retransmission of all subsequent messages. This message, so packetized, arrives at a #7 gateway interface and is sent onward to its destination via the #7 CCS network. One of several methods of working could be adopted, two of which are:

(a) Strip off all envelopes and repackage in #7 envelopes. In Figure 9.1b then, INF A becomes INF 1 (and, possibly, INF 2, 3, etc., all lumped together). Such treatment would break the end-to-end X.25 protocol.

(b) Enclose the complete X.25 message within a #7 envelope. Then in Figure 9.1b, INF A becomes all of the PSS initial message (and possibly followed immediately by complete, subsequent, PSS messages).

In the latter case the #7 error control is exercised upon much longer messages, including the X.25 check word, so that the #7 information-carrying capacity is reduced not only by the redundant data, but also by retransmission of redundant data found in error plus subsequent data sent across the PSS network. In both cases the PSS/#7 interface is faced with the problem of reassembling packets, including retransmitted good packets, in correct sequence. The problems of retransmitted, good information was one of the problems that the #7 recovery mechanism was intended to solve.

Is it unfair to demonstrate a number of problems using a scenario that is not, at present, proposed? I could, for example, have chosen a more realistic scenario in which the PSS network uses the PSTN purely as a circuit-switched transport mechanism. The PSTN must, in this case, examine the address to set up the connection and cannot be easily informed of the point at which the connection can be discontinued. Error recovery is purely PSS protocol and destination PSTN terminations must be competent to handle PSS protocols. Non-PSS compatible terminals either cannot be connected or have no error control facilities for, at least, the PSS portion of the connection.

By using a (presently) unreal example, a number of problems, not all of them confined to error control, associated with providing integrated facilities over dissimilar networks have been demonstrated. The easy solution is to suggest that the integrated network will be one network and that all others will decrease and, finally, disappear.

A second aspect of error control concerns the nature of the information being carried. At the extremes are digitally encoded voice information and fully encoded data. According to one measure[2] 40 errors per hour for voice and 18 errors per hour for data are satisfactory limits. If voice and data are integrated over the same network then the more onerous condition for data must be imposed on all communications, or different standards could be used provided the nature of the message can be distinguished. The V5 standards, referred to in Chapter 5, do in fact include performance grading of messages to enable the local exchange to accept or reject a call based on the call type and the current error performance of the local line.

9.4 Performance estimation and measurement

In November 1986 dealing ceased on the floor of the London Stock Exchange and was transferred to computer-based systems, allowing deals to be struck from the office. The event became known as the "Big Bang", and during the first few days of operation was, apparently, more of a whimper as the computer systems struggled to carry a load far in excess of expectations. This cautionary tale serves as an introduction to the subject because in both computer-sizing and network-sizing exercises it is equally crucial to know the nature of the data to be handled. The Stock Exchange estimates were at fault, at least in part, because the complexity of the system options made the task of estimating loadings impossibly difficult.

To illustrate the communications problem Figure 9.2 shows a notional transfer of information, 400 bytes in size, by X.25 protocol in ten packets of 40 bytes each and by #7 in one continuous stream. In the X.25 case an additional 102 bytes of information is added because of the protocol and for #7 an additional 116 bytes is required. The preparation of Figure 9.2 from the figures in earlier chapters referenced in Figure 9.2[2] has involved numerous assumptions, so the actual protocol content of any real message is largely indeterminate. In the case of X.25 the packet size chosen will also have an effect.

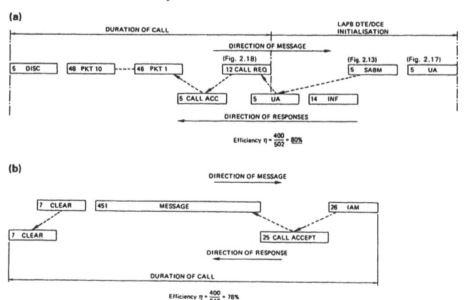

Figure 9.2 Notional message transfer scenarios. Numbers indicate the message portion length in bytes. (a) X.25 message. (b) #7 message.

---

2. Some of these figures appeared in the first edition but do not appear in this edition. The reader will have to rely on the author's integrity with the numbers or check by reference to the base standards.

The problem then is that the use of a layered, enveloping protocol has made the task of predicting the data throughput required to carry a known quantity of raw data more and more difficult because the data is progressively disguised as each layer, or envelope, of abstraction is added with its attendant options and uncertainties.

The network can be measured for bits/s throughput, and the communications terminals can be measured for bits/s output (or reception), but the relation between the two is disguised by the communications protocols.

In Figure 9.2 no allowance has been made for transmission delays through the network components and no allowance has been made for error control activities. Both of these add complications to the calculations but, in general, calculable complications.

For the network, ITU-T provides overall figures for performance for international circuit-switched data connections[124]3 that are summarized here:

	Probability
Call request rejection because of congestion	9×10^{-3}
Established call cleared because of congestion	9×10^{-6}
Established call reset because of congestion	2.5×10^{-5}

From these limits the quantity of data that the network can carry and still meet the limits can be established, but the task of converting this into the quantity of actual information, excluding all protocol elements, is very largely speculative.

9.5 System complexity

9.5.1 In signalling

The previous section on performance estimation and measurement leads naturally to a far more speculative discussion of the desirability of systems using complex and sophisticated protocols.

It must have been apparent to the reader struggling with the contents of Chapters 2, 3 and 4, not to mention Chapters 6 and 7 that, if understanding of the concepts and protocols discussed is difficult, as it undeniably is, then implementation in components and networks must be equally complex. If any doubt remains, refer back to Figure 4.6 and think, for a moment, of the complexity of the systems required to implement the #7 MTP basic error control.

It has been an article of faith among communications professionals since SPC was introduced to telephone exchanges that features implemented in soft-

---

3. The figures quoted are actually from the 1984 version. The current figures are equivalent but more complex so that the earlier version is more suitable for the illustration.

ware could be considerably more complex than features realized by a hard-ware design. This may well have been a misconception bred of an uncritical acceptance of methods well suited to "off-line" computation activities and application of these methods to real-time systems with questionable success.

The development history of all the SPC switching systems now in use worldwide and making the ISDN possible has been a history of massive engineering teams, writing, rewriting, patching, discarding and writing again extensive libraries of software. The theory of complex software systems is only economical in fact if the software can be written and used with a minimum of debugging activity. This has not, in practice, proved to be possible despite the development of very sophisticated software editing, test-bed and qualification tools.

It is significant to this argument that the present generation of digital SPC switching systems, most conceived in the 1970s, has been upgraded and presented as the switching systems for the ISDN until the millennium and beyond. No manufacturer appears to be actively developing a system for the next generation, and several manufacturing enterprises have merged or disappeared as a result of the expense of developing the present generation.

Indeed, with the development of SDH and ATM, it is now unlikely that there will ever be a new development of a main public exchange switching system as we now know them. Future developments will more likely be of devices such as intelligent add-drop multiplexers and digital cross-connect systems allied with the existing trend to put more and more of the complexity of the system into the apparatus at the user's premises.

Against this background, is it not relevant to question the use of protocols and layered models that compel the network components to remain at their present level of sophistication or to "progress" to even greater levels of complexity? The concept of OSI is excellent in compelling interworking between "open" systems performing off-line tasks and interconnected by dedicated links. The OSI concept, extended to "open" systems communicating via "open" public networks in real time, is, at the least, questionable.

In previous chapters the need for a multiplicity of similar CCS signalling systems in different parts of the network has been questioned. It has been accepted that CCS itself is a "good thing". Indeed, the excellence of CCS as an economic medium for providing secure signalling for a multiplicity of traffic circuits is undoubted. Further, CCS can be an excellent means of concentrating network processing power in just a few locations. But this is a trend that goes directly against the trend to place greater complexity at the user terminal. These advantages of CCS are all hardware advantages. In the end, it is only economic to reap the benefit of these hardware advantages if the cost of software development can be recovered from the price of the (reduced) quantity of hardware supplied.

9.5.2 In message switching or circuit switching

It is a historical accident that the ISDN proposal has been preceded by implementation of message-switching networks for data. It is not altogether clear to the author what real advantage accrues from using packet-switching techniques for the public data network rather than circuit switching. Perhaps a major advantage of packet switching is that it enables the service to be measured and charged according to throughput rather than according to permanent circuit provision.

The PSS, as it is realized using X.25 protocols, has numerous advantages and is, on the whole, user friendly. There are, it is true, significant differences between the packet-switched services as they are at present offered by various national administrations. Some leave a good deal to be desired. As the PSS exists, and as it conforms, in general, with OSI (or vice versa?) it must be accommodated by the ISDN as a co-operating and interconnecting network. The effect that the PSS recommendations of CCITT (X.3, X.25, X.28, X.29, X.75, etc.) have had upon the I-series recommendations is self-evident. Not much, if any, of this influence is obviously malign, but a true judgement is difficult to make. It is, at least, arguable that an ISDN conceived in the absence of packet-switching influence might have been easier to implement.

9.6 Clear channel capability (bit sequence independence)

We now come to a problem that is peculiar to the "24-channel" PCM world of North America and the Far East. The problem was discussed in the first edition but is included here, despite the fact that it is no longer a live issue because of the introduction to the problems of encoding laws that have a continuing relevance. It is a problem that is, in fact, caused not by the choice of 24 channels rather than 32, but by the particular implementations of 24-channel multiplexing and line coding that have been adopted. Of prime importance for the 24-channel world, it is of considerable interest to the rest of us as a cautionary tale on how good decisions can become questionable to disastrous because of the progress of technology. The current desire to perform tasks with PCM transmission never envisaged by the original system designers has posed problems of substantial proportions.

When digitally encoded voice signals are being transmitted it does not matter much if the encoded signal is provided in a 7-bit word or in 8 bits or even a mixture of the two. The original PCM system, T1/D1, used 7-bit encoding in an 8-bit word, the remaining bit being reserved for channel-associated signalling. A 7-bit word was, however, inadequate for transcontinental transit connections with, possibly, eight transit switching centres and several analogue-to-digital conversions en route. The current 24-channel system there-

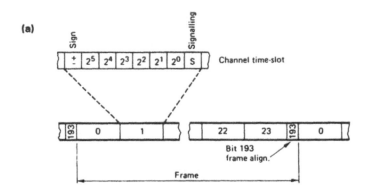

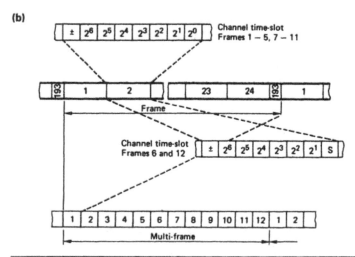

Frame no.	Frame alignment Bit 193	Multiframe alignment Bit 193	Bit number in each channel time-slot		PCM signalling channel
			For voice	For signalling	
1	1		1–8		
2		0	1–8		
3	0		1–8		
4		0	1–8		
5	1		1–8		
6		1	1–7	8	A
7	0		1–8		
8		1	1–8		
9	1		1–8		
10			1–8		
11	0		1–8		
12		0	1–7	8	B

Figure 9.3 (a) Bell D1 24-channel PCM system. (b) Bell D2 24-channel PCM system (Recommendation G.733).

200

(c)

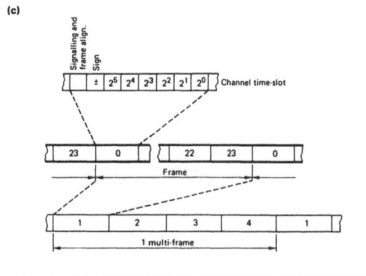

Frame number	Allocation of Bit 1
1	Signalling for individual channels
2	Spare
3	Signalling for individual channels
4	Frame and multi-frame alignment

Multiframe alignment pattern
Bit 1,
channel x
of frame 4

0	1	2	3	4	5	6	7	8	9	10	11	12	13	14	15	16	17	18	19	20	21	22	23
Value																							
1	1	0	1	0	1	0	1	0	1	0	1	0	1	0	1	0	0	0	0	0	0	0	0

Figure 9.3   (c) UK 24-channel PCM system.

fore uses an 8-bit word but "steals" the eighth bit every six frames for a channel-associated signalling channel of lower capacity than the T1/D1 system. Figure 9.3 illustrates all four PCM transmission systems ever used in public service (described in reference 2), the UK 24-channel system being included to show that the UK had a similar potential problem that does not exist in practice because there was never any intention to use UK 24-channel links for anything other than voice transmission.

The problem then is that all systems, except the 30-channel system, offer only 56 kbit/s clear channel capacity for data. Any attempt to send 64 kbit/s data is obstructed by the permanent or occasional absence of the eighth bit.

This problem, as stated, is not particularly difficult to solve. Any use of data as part of the ISDN is intended to be used in conjunction with CCS, and use of CCS eliminates the need for channel-associated signalling. Remove the signalling bit facility from T1/D1 and D2 and we have 64 kbit/s clear channel capability.

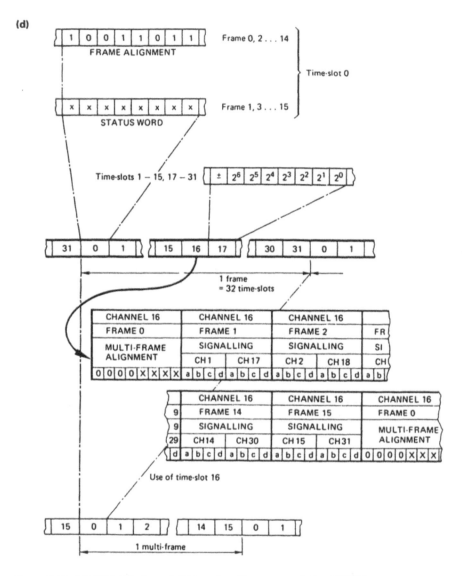

Figure 9.3 (d) CEPT 30-channel PCM system (Recommendation G.732).

The real problem becomes apparent from a glance at Figure 9.4. In attempting to achieve a data stream with a reasonably limited bandwidth and, particularly, to exclude DC components from the line signal, a tertiary line signalling code is used by all PCM transmission systems. The 24-channel world uses alternate mark inversion (AMI) (Fig 9.4d). The 30-channel world goes further and uses high-density binary modulo 3 (HDB3) (Fig. 9.4e). HDB3 limits consecutive zeros to three, whereas AMI, used in conjunction with particular multiplex arrangements, limits consecutive zeros to seven.

202

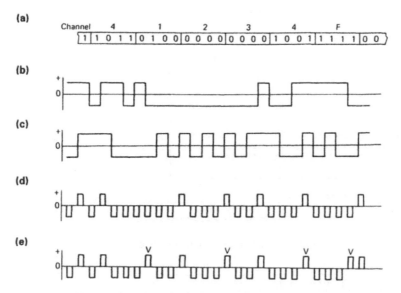

Figure 9.4 PCM line coding. (a) Binary bit stream (part of Fig. 3.6). (b) Symmetrical mark/space telegraph waveform. (c) Alternate digit inversion. (d) Alternate mark inversion. (e) High-density binary modulus 3 (HDB3) (V, violation; fourth zero inverted).

The simple-minded regenerators in the 24-channel transmission systems can, and do, therefore look for strings of zeros greater than the absolute limit of 15 and generate an alarm signal if such strings are detected.

If 64 kbit/s data is being transmitted there is, of course, no control over the information content, which may validly contain long strings of zeros. The HDB3 line code overcomes this problem by, in the last resort, introducing bipolar violations of the AMI rule, inverting the fourth zero of a string. Such a bipolar violation method could solve the clear channel capability problem and such a scheme, bipolar with eight-zero substitution (B8ZS), has indeed been proposed. Figure 9.5 illustrates B8ZS. There remains a real problem however. Existing multiplex equipment is designed to detect and remove bipolar violations, thus preventing the violations from being passed transparently from end to end of the network. Figure 9.6 illustrates the action of such a multiplex upon a B8ZS signal. Further, existing regenerator equipment detects violations as alarms and actually uses particular patterns of violations as test signals.

To introduce B8ZS into the 24-channel network will require detailed examination of the components of each and every link destined to carry 64 kbit/s data and extensive retrofitting of all such links. The ISDN must include features for recognizing the need for clear channel capability and categorizing possible routings with this end in view.

There is an alternative solution to overcome the problem that obviates the need for categorizing network links and retrofitting network elements at the expense of increased complexity, storage and delay in the terminating equip-

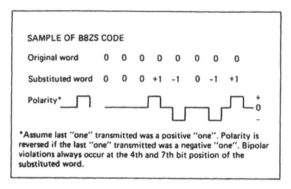

Current CCITT recommendation to obtain 64 CCC
      – Removes zero octets and substitutes B8ZS code
      – B8ZS code contains bipolar violations in 4th and 7th bit position.

Figure 9.5 Bipolar with eight-zero substitution (B8ZS) coding.

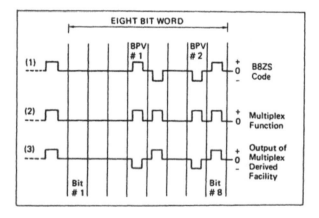

Figure 9.6 B8ZS through digital multiplex

ments. This is zero-byte time slot interchange (ZBTSI). This method is illustrated in Figure 9.7, which shows a four-frame example with zero words in channels 3 and 25 of the 96 channels. The multiframe alignment bit is "borrowed" and inverted if zero channel words exist. The zero channels are removed and replaced by "empty" channels at the start of the four-frame period, and these empty channels are filled with a word indicating the position of the zero words. Thus, at the cost of, perhaps, delaying slightly the process of frame alignment, the possibility of zero words in the line signal is removed without any requirement to replace network components.

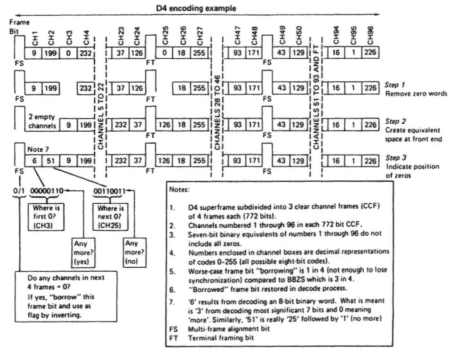

Figure 9.7 ZBTSI signal-processing method.

9.7 Network numbering

The introduction of direct dialling in (DDI) to PABXs, among other innovations, introduced the potential for problems in a shortage of subscriber numbers in existing network numbering schemes. An early UK solution to the problem was to cease to include a letter significance of the exchange name in the number. At about the same time that the UK effected that change, the Americans introduced area codes as well as exchange identities. In America and Canada at the moment one dials ten digits: xxx for the area, xxx for the exchange and xxxx for the subscriber. The exchange significance is notional as the typical size of an American urban exchange may exceed 50000 lines and would therefore use several exchange codes.

In both the UK and the North American numbering plan, although the alphabetic significance of the numbers is not used, there is still a geographical significance to the number. "117" in England indicates numbers in Bristol city.

Network numbering is standardized, to an extent, by the ITU-T. Figure 9.8 shows a synopsis of the recommendations defining international numbering. ITU-T Recommendation I.330[125] (not shown in the diagram) describes ISDN numbering and addressing principles. Recommendation E.164[126] describes

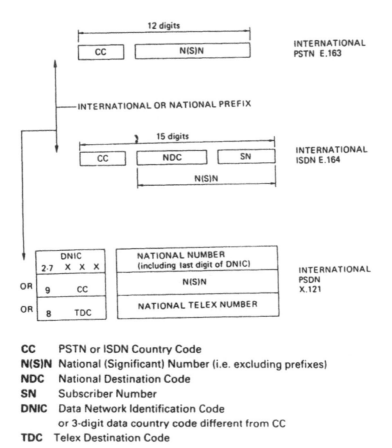

CC   PSTN or ISDN Country Code
N(S)N National (Significant) Number (i.e. excluding prefixes)
NDC   National Destination Code
SN   Subscriber Number
DNIC   Data Network Identification Code
       or 3-digit data country code different from CC
TDC   Telex Destination Code

Figure 9.8 International network numbering.

the numbering plan for the ISDN era and identifies the need for interworking arrangements between ISDN and present dedicated networks. Recommendation E.165[127] sets a specific time, 31 December 1996 at 23.59 hours co-ordinated universal time (UTC), after which all ISDNs and PSTNs can use the full capability of Recommendation E.164.

The ISDN basic access can accommodate up to eight terminal devices. The ISDN concept includes the possibility of a user being identified at any point in the network. Existing development such as DDI to PABXs and Centrex will be supported and encouraged by the ISDN. All these aspects will cause an explosive growth in the requirement for unique network numbers. At present in the UK, subscribers to Mercury must have a different directory number to that required for the BT network. Thus, an organization with PABX exchange lines incoming from both BT and Mercury must have a different directory number for each group. Given adequate facilities in the SPC local exchange for number interpretation, there is no technical reason why this should be so.

All these considerations lead to a requirement that the public exchanges in the modern, liberalized, network incorporating ISDN must provide facilities for examination of all, or most, digits of the network number. The national and international numbering schemes must be altered as necessary to provide for an adequate supply of unique numbers. In the early 1990s the UK implemented a plan for the London area numbering that replaced the "(0)1" prefix with "(0)71" and "(0)81" and then, in 1995, implemented a change for the whole country introducing an additional digit at the start of the number. All existing numbers now commence with an additional "1" so London has become 0171 and 0181 and so on.

National and world numbering schemes have become an important human resource, and it is of concern to all of us that this resource is conserved. Not all recent decisions give confidence that the authorities are fully aware of their responsibilities in this regard. A very old telephone number, 01 368 (ENTer-prise) 1234, the number of STC New Southgate, now Nortel Europe, used as an example in the "bible" of UK telephone engineers, Atkinson's *Telephony*, has recently been changed first to 081 945 4000 then to 0181 945 4000. This is because STC has chosen to use Centrex on the alternative MCL network. There is no good technical reason why the same number should not be used for the same termination on either network, but the allocation of a different exchange code, 945, to the MCL local exchange avoids the need for existing BT exchanges to examine the subscriber number. This unnecessary proliferation of codes is, however, a misuse of the national numbering resource.

The most recent changes in both North America and the UK, the latter referred to above, do therefore give cause for concern. Have they paid sufficient attention to the need to conserve this national resource? In the UK, OFTEL commissioned a study and published the results[128]. The most economical and sensible suggestion was a gradual loss of the geographical significance of the area code allied to a commitment to extend the numbering scheme to ten digits in the future in a gradual manner. The option chosen and implemented in April 1995 has been to add an extra digit to the front of all numbers, permitting the initial digit to indicate a specific network or service. This abandons any attempt to rationalize the present proliferation of unnecessary numbers for different networks and services and adheres to a geographical significance that is irrelevant in the presence of the personal portable numbering available with the ISDN.

The North American approach is perhaps more sensible, although there too there is no attempt to combine numbering across networks and services. There the area code is being deprived of its geographical significance in a gradual manner and there is a plan for the introduction of a fourth digit to the area code.

In the UK the inevitable result of this ill-considered approach is that we still have a problem. The latest in the series of proposals and counter-proposals, largely covering ground treated in the original OFTEL report, has only recently been published[129].

9.8 Centrex and virtual private networks

Modern networks incorporating intelligent network capabilities and ISDN and B-ISDN access possibilities can allow the user more ability to manage the network, both the user's own private network and the user's access to public networks and services. The intelligent network and the introduction of ISDN have also brought back into prominence the concepts of Centrex and virtual private networks (VPNs).

Centrex is clearly a feature that can be more easily provided by the ISDN, but its importance is greater than this. The next chapter will discuss, among other things, the putative attractiveness of the ISDN to the subscriber, and it is in making the ISDN attractive to the subscriber that Centrex may have an important part to play. At this stage we need to introduce Centrex, what it is and how it may be provided by the ISDN. To do this requires a short excursion into private telephony, the PABX.

The principal and traditional function of the PABX is to provide a central answering service for the organization. The original switchboard was manual and provided incidental savings in local line plant by concentrating traffic from extensions to exchange lines. In the mid-1950s, the very latest thing in PABX technology was the cordless board, which saved the operator from permanent involvement with the call to see when it had finished and, by running three wires to each telephone and fitting a recall button, greatly enhanced the features available to extensions. The principal new feature introduced by the cordless board was enquiry and transfer.

Provision of a PABX very largely duplicates facilities already available to extension users directly connected to the local exchange. As the local exchange has become increasingly sophisticated with the advent of SPC this has become more and more true. There is nothing now to prevent the PABX operator features and all current PABX features, which now include short code dialling diversion, conference, etc., from being supported by the SPC local exchange.

AT&T appreciated this and offered Centrex, a centralized private exchange service, on its second SPC development, ESS no. 2 in the mid-1960s. ESS no. 2 was developed for rural and suburban applications where the economics of providing the local network was little different to wiring up the workplace and the increase in lines to the exchange, one per extension rather than one per exchange line carrying concentrated traffic, made little difference to the costs. In urban environments where local distribution is much more costly, the economics of Centrex, in the absence of "pair gain" systems, is more questionable.

With the exception of AT&T, the policy of administrations has consistently been to structure their systems and their tariffs so as to preclude the provision of Centrex. The system arguments and the economics are, however, evenly balanced even in an analogue, non-multiplexed, local network environment.

The AT&T Centrex innovation was not well received, and little penetration of Centrex into the American network occurred in the early years after its introduction.

When the local network uses a digital multiplex and the local exchange is digital SPC, then the economics of providing Centrex definitely swings in its favour, even in the urban environment. Despite this, the marketing case for Centrex to an administration with PSTN monopoly, and also, perhaps, a monopoly in PABX supply, is still weak. It required the impetus of deregulation and liberalization to alter the picture as perceived by the administration in favour of Centrex.

The liberalized market, however, increases the utility of Centrex to the network provider. It provides a means of retaining the market for private communications that might otherwise be lost to other suppliers. This fact has not been lost on the regulatory authorities, and both the FCC in the US and OFTEL in the UK have had to reconsider their attitude to Centrex. Does it constitute basic or enhanced service? Is it a bearer service or a teleservice in ITU-T terms? Where is the interface to the public network? Just at the time that Centrex has come into its own with the network providers its existence is called in question by the regulatory authorities.

Figure 9.9 illustrates Centrex in the ISDN environment. The digital multiplex has reduced the cost of providing a voice channel for each extension. CCS is now under network provider control and therefore without the reservations attendant upon subscriber access to the network signalling. CCS also provides all the sophisticated PABX features, as indeed ISDN does, from the local exchange. The use of permanent and semipermanent connection options allows the creation of a VPN in a far more efficient fashion. For example, links dedicated to a 9 till 5 office can be reused for the evening PSTN busy hour. Links dedicated to voice by day can be used for bulk data downloading by night.

Centrex then is the provision of private communications switching functions by means of public facilities located in the public telecommunications network. The Centrex service will be more conveniently provided by the local exchange, as shown in Figure 9.9, but in the early stages of implementation it can be resident in a special exchange remote from the user. This latter arrangement is that used in the initial Mercury 2110 service in the UK and in that originally proposed, but not offered, by BT.

It is not necessary to talk of virtual networks in terms of Centrex, but it is certainly easier to do so. In switching, particularly digital switching, it has become convenient to refer to a permanently or semipermanently held "connection" through the switch as a virtual circuit. Such a connection could be a

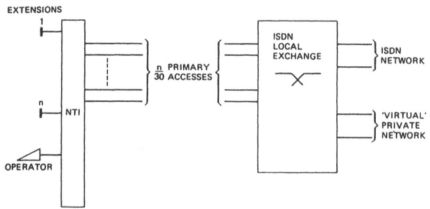

Figure 9.9   ISDN Centrex service.

permanent path between a terminal and a register or such-like device.

When the user rents a private circuit from the operating company, what is rented is the assurance of permanent connection between two terminals. The user certainly does not rent a particular combination of lines and channels; the only parts of the connection that are permanently defined are likely to be the local ends. Between these the operator can, and does, vary the connection path. The operator may, however, have to declare to the user certain characteristics of the line and therefore can only vary the routing of the line within the declared parameters. To this extent, therefore, every private circuit is virtual. In digital working, of course, the connection is re-established for every time slot and hence the use of the term "virtual" came into being to describe a virtual permanency.

For a virtual private circuit the proposal is to market a technological fact and offer not a permanent guarantee of connection terminal to terminal but an assurance of connection when this is required. The user performs the same actions as would be necessary to seize a private circuit, but the connection is made on request out of the full stock of options available to the public network.

To the same extent that Centrex provides a virtual private switch, the VPN interconnects the Centrex "nodes" so that the user finds the difference between the virtual and the real indistinguishable. Clearly there can be a real and significant difference in that the virtual approach can be far more economic and tariffed accordingly.

Provision of Centrex and virtual private networking leads to virtual network management. Perhaps the greatest good to the user to be expected from the combination of Centrex and virtual networks is the ability to manage the private network from a terminal. Just as Centrex allows a private operator to deal with staff and the public as if this were a private exchange, the extension to a virtual network allows the user to effect changes electronically. Within

210

organizational limits, set by the network operator, the user can add and change facilities, terminals, private links, etc., and can obtain full and immediate details of the effects, in costs and traffic, of such changes.

One example of a VPN is illustrated in Figure 9.10. The network includes automated factory locations, office locations using a considerable amount of IT and a mainframe computer serving the industrial group. In such an arrangement there may well be interfacing problems in feeding information from (say) production to office terminal systems or to the mainframe data banks. The presence of CCS from terminal to terminal (ITU-T DSS 1 or, in the UK, DPNSS) makes it possible for the terminal to define the virtual connection. Such a connection could, if necessary, include protocol conversion features provided as an external network service, such as those shown in Figure 9.10, or as an internal service of the organization.

The combination of the ISDN basic access, representing, as it does, a small combined PABX and LAN, with the concept of the VPN is very powerful indeed. Organizations could operate as if there was but one facility interconnecting all their main locations and their high street branches without the need, in most cases, to set up their own private network.

All that has been said may well be called into question by recent European Union directives seeking to impose open network provision (ONP) on the public networks within the EU. The objective is to ensure that all public network operators offer a basic minimum of typical services. One of these that differs widely within the Community at present is the provision of leased lines[130]. What needs to be said here is that the European standards now being introduced to mandate a common minimum provision of ONP leased lines throughout Europe ought to take account of the possibility of the leased line provision being made through the means of VPNs. At present there is no real assurance that this will be the case.

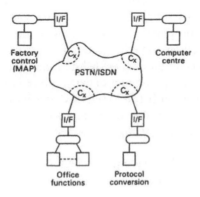

Figure 9.10 Virtual private network for data.

9.8.1 Centrex in the UK

The second network provider in the UK, Mercury Communications Ltd. (MCL) introduced Centrex to the London area in 1989. At about the same time there were active rumours that BT were about to follow suit. MCL utilized at first a single DMS 100[4] local exchange. The first customers, who, necessarily had to be rather large, were provided with a DMS 100 concentrator unit on their premises. They cannot have seen much difference in the quantity of equipment on site between Centrex and having a PABX. Within two years MCL acquired the largest UK PABX rental company, Telephone Rentals Ltd and MCLs' marketing activities for Centrex were very much scaled down as a result. They did acquire one very big customer, Standard Telephones and Cables Ltd (STC) who today are part of Nortel.

In the event BT joined the market some five years later, and they too chose to implement Centrex[5] using a seperate network consisting of interconnected DMS 100 exchanges. Each local exchange hosting Centrex subscribers is equipped with a DMS 100 Small Remote Unit (SRU). In the UK the public network is almost entirely re-equipped with digital SPC exchanges of either the Ericsson AXE 10 system (called System Y) or the home produced System X. It appears that neither of these systems provides Centrex features to the standard of functionality of DMS 100. It is true however that BT performed one of their first Centrex trials in Bristol based upon an AXE 10 exchange[149].

The dependence of both MCL and BT on what is, in effect, a separate network for Centrex and VPN is of course at the least a pity, more properly, a network designer's nightmare.

9.9 Switched multimegabit data services (SMDS)

Telecommunications has not been noted for sparing the public from the need to learn new words, most of them made up of initials. Switched multimegabit data services (SMDS) is one that does not represent a new technology so much as a marketing person's bright idea for selling something already in stock. Having successfully offered to the public 64 kbit/s and 2.048 Mbit/s links there is still a need for yet higher capacities for the large volumes of data required by some major private network users. Even so, it is difficult to sell this capacity in the digital hierarchy units of 8 and 32 Mbit/s. SMDS is an attempt to provide the user with the equivalent capacity by means of digital connections not unlike those envisaged in the VPN.

Having started from a concept in the USA of providing switched data serv-

---

4. The DMS 100 system is manufactured by Northern Telecom, who changed their name to Nortel Ltd in 1995.
5. BT Centrex and VPN is called Featurenet.

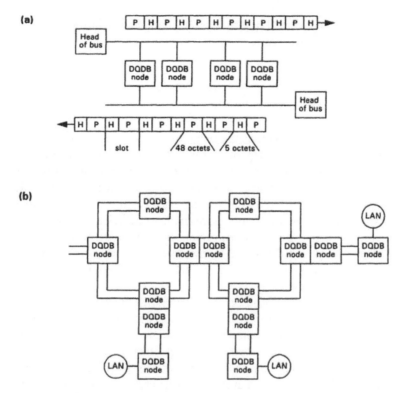

Figure 9.11 Distributed queue dual bus architecture and networks. (a) Distributed queue dual bus subnetwork. (b) DQDB in a public data network.

ices on demand and in the capacity demanded, SMDS has developed into a concept of a fast PSS not unlike frame relay but with built-in security and very similar indeed to ATM. This maturing of the concept has brought with it a real technological advance in the form of the protocol most often envisaged with SMDS, the protocol, or interface technology, known as distributed queue dual bus (DQDB). This is defined in the American metropolitan area network (MAN) standard IEE 802.6,[6] which represents an alternative method of achieving metropolitan area B-ISDN capability.

Each terminal on a DQDB bus (Fig. 9.11) is required to count empty packets so that enough packets for waiting information get to the end of the bus before the terminal seizes a packet for its own use. The packets are identical to those used in ATM. Whole networks can be constructed from DQDB buses, as is illustrated in Figure 9.11.

SMDS services are widespread in the USA and are appearing in the UK. The prospect of SMDS being freely available calls into question the advisability of choosing options such as frame relay and even perhaps has endangered the

---

6. The IEEE 802 family of standards defines data network protocols including Ethernet and FDDI LAN protocols [131].

introduction of VPN, which is at present mainly available on a limited scale for international connections.

9.10 Network and system reliability

Modern systems enjoy a much increased reliability. It is perhaps fortunate that the systems also introduce the capability for remote operation and maintenance because this has become essential to keep the maintenance force in training for their task. When failures occur so infrequently and, when they do occur, they can often be solved by the simple replacement of a unit, there is very little happening to give the maintenance engineer sufficient experience of failure modes. Thus, it is not only convenient but necessary to extend the maintenance area to cover sufficient equipment so that there is enough going wrong, even with modern reliable systems, for the engineer to learn from experience of actual failures.

Another aspect of this increased reliability is that the systems are realized very largely by means of software modules. The software itself will have been exhaustively tested during the manufacturing phase to exclude all but the most abstruse errors. There is, however, a danger that a route through the software process is so rarely taken that it has not been tested in the software emulation phases. When it does happen in practice, and goes wrong, the maintenance engineer is faced with a completely unexpected new occurrence whose source may be very difficult to trace or which may be difficult to reproduce in order to study the system behaviour.

The US has been using a CCS network interconnecting all its SPC telephone exchanges for many years. The network is illustrated in Figure 9.12, which shows the considerable degree of back-up CCS links used to ensure that the network will be always available. Despite this provision for network security, there have been several failures of the CCS network in the past few years, two of which were sufficiently serious to deprive much of the country of telecommunications for most of a working day.

One of these faults occurred when a new release of CCS software had been installed in some or all of the exchanges and a software fault caused one exchange to fail. Because this exchange broadcast the fact that it had failed as a *link* failure, without any further fault, surrounding exchanges concluded, wrongly, from this failure that they too were in error and announced their failure. The network therefore progressively collapsed. Paradoxically, the situation would not have occurred in networks, such as the CCS network in the UK, that use a simpler back-up interconnecting pattern not so heavily interconnected to avoid error.

This is a cautionary tale to underline the need for network managers to validate the networking and the nodes, an activity that is essential if their net-

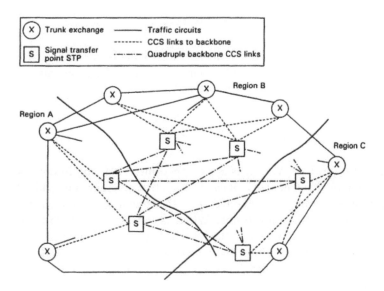

Figure 9.12 US quasi-associated CCS architecture.

works are to remain operational. They must also be very careful to specify and supervise the introduction of new features and facilities that, if in error, may have a disastrous effect on their networks.

A second tale is much more mundane, but just as important and, again, is an American experience. To save on the costs of energy, it has become the practice in the US for operators to switch the exchange power supplies over from the mains to their own back-up generators. It is only in an economy such as America's, where oil fuel prices are subsidized, that such a method of saving could be contemplated. In late summer 1991 a main New York switching centre performed this automatic switchover for the working day to the directions of a time clock setting. All the members of the maintenance force were away on a training course and nobody noticed that the generator did not start. A number of exchanges in the building, including transit exchanges carrying air traffic control traffic between the three main airports, therefore continued to operate for some five hours until their stand-by batteries discharged and much of New York and all air traffic came to a halt as a result.

Network management at its most basic can sometimes fail; the manager has to look for both the most abstruse and the glaringly obvious. Reliable networks[132] are a function not just of reliable component systems, but of their systematic assembly into reliable networks and of eternal vigilance on the part of the network manager, vigilance extending from the design of the network through network events to the whereabouts of the network technicians.

9.11 B-ISDN issues

Until now, the chapter has dealt with issues that were equally relevant to ISDN or to B-ISDN. It will conclude with an account of some of the problems and solutions that are relevant to the B-ISDN alone.

9.11.1 ATM over SDH links

Figure 9.13 illustrates how ATM cells are mapped into SDH, continuing the descriptions contained in Chapter 7. It can be seen that no attempt is made to contain the ATM cell within the SDH envelope – the stream continues without interruption into the succeeding frame. Whereas synchronous streams are contained in tributaries and have a pointer mechanism indicating where they are, the ATM stream is fixed within the envelope and must be treated as a whole. The resulting problem, briefly described here, is well described and illustrated in reference 133.

A common way of providing SDH transmission links will be to provide SDH rings as shown in Figure 7.2 and with similar properties to the FDDI ring architectures discussed in Chapter 2 and illustrated in Figure 2.9. While the ring is unbroken a number of add-drop terminals contributing ATM streams to the total transmission will be accommodated, but once a fault occurs, breaking the ring, then ATM cells from some terminals approach the serving node from one direction while those from the remaining terminals approach from the opposite direction and the serving node has no way of knowing which stream comes from which terminal.

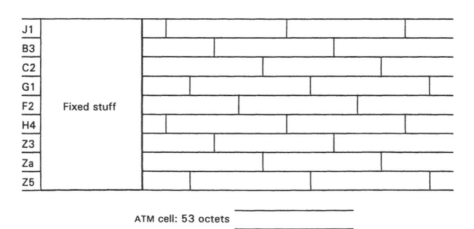

ATM cell: 53 octets

Figure 9.13   Mapping of ATM cells in VC-4-Xc.

Clearly, there is more work for the ITU-T to do here. Recommendation G.709[97] indicates that ATM mappings are to be studied further. At present, the solutions are to provide STM-1 level service to each user or to abandon any thought of combining ATM over SDH.

9.11.2 ATM flow control[134]

Chapter 6 introduced the reader to the deliberations of the ATM Forum. This body has been, more than the ITU-T, concerned with the performance of ATM in both LAN and wide area network environments. In the wide area particularly there will most probably be a mixture of constant bit rate (CBR), variable bit rate (VBR) and available bit rate (ABR) traffic. The first two are controlled by agreement between the network provider and the user on the bit rate to be used, but ABR, by definition, uses a bit rate determined by the traffic level on the links, using up the remaining link bandwidth after the other services have been accommodated.

This requirement involves the provision of a feedback loop from the terminating user and the network that will indicate to the sending end the rate at which the ABR traffic should be transmitted. Two competing methods have been discussed for providing this flow controlling feedback:

(a) Rate-based flow control. The destination indicates the rate at which it is able to receive information in a returned resource management (RM) cell and the network is free to modify this RM indication according to the available bandwidth.

(b) Credit-based flow control. The destination and the intermediate switches return credit indications in the RM cell indicating how much capacity for buffering messages is available. The sending end may only transmit up to the limit of the received credit.

In rate-based systems the sending station will transmit a request to transmit including the rate of transmission desired and will only commence transmission on receiving the returned RM message indicating the rate to be used. Transmission will continue but be modified as necessary to the rates indicated in the returning RM cells. Thus, all the necessary complication is localized in the terminals and the approach is therefore that preferred by the public network providers for the wide area networks.

In credit-based systems a further piece of information is contained in the RM. This indicates the availability of buffers at each node in the virtual connection. This allows much more immediate variation in the rate of transmission and a much better approach to full utilization of the available bandwidth. But the buffering must be sufficient to contain all the data stream that would be sent during one round trip's delay. In wide area networks (international connections) this could be considerable, but the resulting control is much finer and permits data to be passed over the ATM network with little or no difference to that experienced on other conventional LANs (Ethernet, FDDI, etc.).

For the LAN therefore, credit control is the preferred option whereas for the wide area network, the public networks, rate control is preferred. At present, the standards allow for rate control but the discussions continue.

9.11.3 SDH pointer problems

In moving from analogue transmission of information to digital transmission many of the traditional sources of impairment owing to loss and distortion have been removed. However, a new source of impairment has been gained related to changes in the phase of the digital signal because of changes in the speed of transmission. Such impairments are called jitter and wander. Jitter is the phase modulation of the data pulses from equipment, which causes the data pulses to arrive either early or late, thus potentially generating transmission errors. When these variations are relatively rapid, say caused by changes in transmission delay owing to digital processing changes, the resulting impairment is known as jitter. When the variations are quite slow with time, caused by transmission variations due to, say, temperature changes, the resulting impairment is known as wander.

The two main parameters used to characterize jitter are the amplitude of the phase modulation, expressed as a peak-to-peak deviation, and the frequency of the deviation. The jitter frequency is typically a combination of the different frequencies generated by the components and equipment involved.

This introduction gives a clue to the problem. SDH, by its nature, introduces changes in phase. Every time there is a change in the pointer mechanism the signal to which the pointer relates changes phase. SDH introduces jitter. When the whole world is SDH this will cease to be a problem. The end-to-end jitter of a digital signal sent via SDH will contain the same jitter as a plesiochronous transmission of the same data over the same medium plus additional jitter caused by the changes in phase dictated by the pointer mechanism to overcome changes in the phase relationship of the tributary signal with the synchronous SDH envelope. This necessary increase in jitter is allowed for in the SDH specifications.

In the interim situation however, while there are SDH "islands", in just the same way as the problems of digital islands as the digital network developed have been discussed, the interconnection of SDH islands by plesiochronous links will cause the jitter introduced by each SDH island to be added. To assess the effect of this additive jitter it is necessary to propose a model of the interim network. Bellcore proposed such a model for SONET, postulating 32 SONET islands, each with ten pointer processing nodes[135]. At the output from the final SDH island the demultiplexed signal has to meet the requirements for signals entering the PDH network as specified in Recommendation G.823[136].

To test that equipment meets these requirements it is necessary to specify test sequences that represent these worst case network conditions. Recommendation G.783[137] describes such pointer test sequences, which aim to emulate the expected effects when the network is in a degraded condition.

One of the tests emulates the condition when the originating terminal within an SDH island loses lock to the system clock. This will cause continuous pointer movements to be generated. In addition to this background activity, additional pointer activity may occur as a result of phase noise from other nodes of the network. This could introduce added or subtracted perturbation to the background sequence every 30 seconds.

The G.783 pattern represents the clock synchronization loss by continuous evenly spaced pointer movements. In practice, however, gaps will be generated in this even pointer movement sequence as a result of the effects of the positioning of the SDH overhead bytes. For example, when a VC-4 is involved there will be a sequence of 87 evenly spaced pointer movements followed by a gap equivalent to three missing pointer movements. Table 9.1 contrasts the performance expected from the G.783 assumption with that postulated by this 87/3 pattern.

Table 9.1 Experimental results showing effect of 87/3 sequence[138].

Pointer sequence	Jitter on demultiplexed DS3 (UI peak-to-peak)
G.783 test: regular pointers with one missing pointer	0.15
87/3 sequence	0.60

Notes: In both cases the pointer spacing was set at 33ms, equivalent to a frequency offset of 4.6ppm. The results were obtained by testing the DS3 drop port from the HP 37704A test set.

The sequences in G.783 will be revised to include this effect. The account illustrates that the design of SDH terminal equipment will have to be a good deal more stringent as a result of this. It illustrates the inevitable problems inherent in mixing SDH with its pointer mechanism, intentionally introducing phase discontinuities into the existing plesiochronous network.

9.11.4 Synchronization in an SDH world

Traditionally we have obtained synchronization at every digital node in the network by locking the local clock to the timing of the bit stream of one of the incoming 2 Mbit/s systems. The discussion of the previous section has alerted us to the fact that, with SDH, this may not be such a good idea.

If the synchronizing 2 Mbit/s system comes to us via SDH then it will have been subjected to phase changes by the pointer mechanism. It is no longer a good guide for timing. It is therefore necessary to ensure that the synchronizing system is never passed via SDH involving permanent retention of a PDH network, for synchronization if for nothing else. This involves network providers in the quite complex work involved in giving assurances that this particular circuit is truly plesiochronous and does indeed never pass over SDH. This would not be practical as a long-term solution.

An alternative would be to develop advanced pointer processing designs that would maintain the impairment of the synchronizing signal within acceptable limits. Such advanced designs are being studied[139].

Alternatively, synchronization can be obtained from the SDH line rate itself. This too is being studied, and guidelines are expected from ITU-T Study Group XIII. It would mean, however, that every node requiring synchronization should terminate an SDH system. Nodes that do not do so would become less well synchronized or would demand SDH-free 2 Mbit/s systems.

The best alternative seems to be a move to synchronization from a source external to the network. This is now possible with the provision of satellite navigation systems, which could provide the master timing source for the whole world, removing our dependency on timing over the network altogether. If SDH pushes development in this direction it will be another feather in the SDH cap.

9.11.5 Optical interconnect

In introducing SDH we started from the provision of digital cross-connect for leased line systems. This has introduced the idea of two sorts of switching node: the cross-connect, switching rather slowly to preprogrammed demands or to less pressing needs; and the telephone exchange, performing circuit switching in real time. For both of these, but particularly for the first, there is now the option to switch not electrically, but optically.

At the centre of the DCC and, eventually, at the centre of the telecommunications exchange one may expect to see an optical switch. It is perhaps too soon to dwell for long on the technology in this edition, although optic switching will be present in DCCs before the third edition is published. A good discussion of an existing field trial is provided in the reference 140.

Chapter summary

This was a chapter full of questions, dealing, as it does, with the technical challenges still unresolved in the emerging ISDN and B-ISDN. The promise of a technical chapter was soon broken by returning to the U interface and updating the account first given in Chapter 3. It seems probable, and logical, that the U interface will at last receive international regulatory recognition.

However, a cautionary tale from the UK illustrated the problem of the regulator in providing enhanced services via the ISDN regardless of where the interface between the subscriber and network is located.

More technically demanding was the problem of error control and the relationship to be adopted between the roughly similar error control mechanisms operating in dissimilar but co-operating networks. There is also the problem of imposing standards of accuracy upon international networks carrying such radically different traffics as voice and data.

The error control discussion leads naturally to a section on the difficulty of measuring throughput in a network passing "enveloped" multilayer protocols. It is seldom possible for the measurement to assess the real data passing as distinct from the combination of data plus, indeterminate, protocol material.

These discussions acted as a reminder of the greater sophistication of the protocols being used and led, at least, to questioning the wisdom of such sophistication. The thesis was introduced that we have developed along a path with milestones as follows:

(a) SPC: software is "easier" than hardware; software allows complication; development delays owing to software errors.

It was suggested that there may be an alternative path:

(b) SPC: design optimum hardware/software balance; develop modular hardware, software and protocol elements.

Clear channel capability is a specifically 24-channel problem but might become a more general problem if, for example, satellite links were to be provided at 32 kbit/s. It is instructive in general because of its demonstration of the engineers' tendency to "get it wrong" when vision is limited to the immediate specification. Possible solutions were indicated and their merits discussed. This discussion gave an opportunity to revise the fundamental PCM protocols.

Less technical, but no less intrusive, are the problems of network numbering. The ISDN introduces many more network addresses, perhaps as many as three for every basic access, and the problems are exacerbated by existing, unresolved differences between the present international numbering plans. No real solutions have been offered, but designers of new systems for the ISDN have been warned to make adequate allowance for development.

Centrex, private switching features provided by the public exchange, is not it would seem immediately relevant to the ISDN until we consider, in the next chapter, the potential inherent in Centrex for introducing subscribers to the ISDN.

Another example of ISDN facilities being provided in other ways in advance of ISDN implementation is the SMDS. This was introduced because it is on offer and adds to confusion as to when, and how, to choose the ISDN. The spread of SDH broadband transmission will, no doubt, limit the period during which SMDS will be a realistic option.

The chapter ends with discussion of four problems and a solution, all specific to the B-ISDN. There is a difficulty in mapping continuous ATM streams into the SDH envelope. There are difficulties in controlling ATM ABR services to use up the remainder of the transmission capacity. The marvellous SDH pointer mechanism introduces problems of increased jitter and calls into question how the network can now be synchronized. And, finally, the B-ISDN generates increased interest in switching at the optic levels.

CHAPTER 10

The network integrated and broadened

It has been made abundantly clear that the problems associated with the ISDN are not all, or even principally, technical. It is true, however, that most of the problems do have a technical dimension. In this chapter it is intended to step back from the detail and look more generally at the implications of the ISDN. By taking the broader view the technical content may be diminished but certainly does not disappear. On a more optimistic note, some examples of the services that the ISDN makes possible will also be discussed.

10.1 Integrated charging

In 1981 the author had reason to display some 200 pages of information on Prestel, BT's viewdata system, the first in the world but then languishing rather, having disappointed early hopes for its widespread popularity. One possible reason for Prestel's poor sales record very soon became apparent.

To access Prestel meant renting a Prestel editing terminal from a TV rental firm, renting a direct line from BT, and having BT fit a second jack socket to connect the terminal. To input Prestel frames required the services of a registered information provider, who paid BT a rent per frame and passed the bill to the user with its mark-up. To construct the frames required paying telephone charges for the call to the Prestel gateway and data quantity and call duration charges to Prestel for the services of the Prestel entry computer. Other Prestel information could also be accessed, of course, for which charges for each frame accessed might be levied.

The task of consolidating the various bills received – some monthly, some quarterly – and from different sources – BT telephones, BT Prestel, information provider, rental company – to determine the final cost was almost impossible.

The present billing pattern for users of the Internet is not quite so bad, BT for the call charges and the Internet server for the access facility, but it is destined to become more complex as more of the information provided is charged for rather than being free as at present.

Exactly this problem of non-integrated charging and billing for integrated services is present in practically all the ISDN proposals. The complication is inherent, by definition. The easy solution is to pass the problem to the subscriber, but this could, and does, kill any interest the subscriber might have in integrated services. ISDN could come to be regarded by the customer as an uncontrollable financial drain.

Prestel is, of course, available over the analogue PSTN. A similar problem predating the ISDN has occurred in America as a result of deregulation. A further aspect of the Computer II determination, not touched on previously, is the ruling actually included in the modified final judgement (MFJ), which divorced the Bell Operating Companies (BOCs) from AT&T. This defines local access and transport areas (LATAs) where provision of local (basic) telecommunications services could be a local monopoly but where the LATA monopoly network provider, often a BOC, must provide choice of access to all long-distance carriers. Hence, even before discussing provision of enhanced service, any call outside the LATA will attract charges from three sources: local network provider, long-distance carrier and remote local network provider. This is the situation in the USA today (although, at the time of writing, it is just about to be changed). Add to this the variety of services of the ISDN and the problem is compounded.

This is not a new problem. The CCITT has contended with the problem of international allocation of charges as one of its principal activities for most of its history. The international solutions have always protected the subscriber from the complication, presenting just the one bill from the originating administration that shows a single composite charge for the call. International apportionment of charges is then a co-operative function between co-operating administrations.

There are two important differences between the tasks of international charge apportionment and of local network charge compilation. International gateway exchanges handle large quantities of concentrated, expensive traffic and can afford to spend significant sums on the complex task of collecting the requisite per-call information. Secondly, the relationship is between co-operating national administrations, who will readily agree to use each other's data. PSTN or ISDN local charge compilation, by contrast, is just as difficult a task performed on smaller quantities of less valuable traffic by organizations that may be in direct competition and, therefore, with less mutual trust. The easy answer is to pass the problem on to the customer on separate bills. The argument that is developing is that such action in the ISDN era will completely demotivate the customer, who will refuse to have anything further to do with such a service.

At present in the liberalized UK network, the problem of apportioning charges between the various networks carrying a particular call, which may be a local CATV network, the BT, Mercury or other trunk network and perhaps a mobile cellular radio network, is solved by the various network providers, urged on by OFTEL, negotiating access agreements among themselves. Thus, at present, subscribers are presented with just one bill by their local network provider.

There is, therefore, an urgent need for agreement, possibly at international as well as national levels, on cost-effective methods of providing consolidated billing for diverse network connection and service offerings. The ISDN technology is inherently capable of providing the necessary features although their provision may not be cheap. In liberalized, deregulated environments it may well be a function of the regulatory bodies to insist that consolidated billing is provided at least for the customers who require it.

Figure 10.1 is a simplified illustration of the problem. Available to the subscriber via the chosen local network are a variety of networks offering routes to other networks and access to a variety of services. Also shown, because it is an example that will be used again, is a service associated with the connection itself. Such a service could be provision of identity data on the caller to the called terminal. All these networks and services have to bill their cost components associated with each connection to the originating local network provider. They must also record their charges for checking against the subsequent receipt of the funds distributed by the local network provider.

Figure 10.2 demonstrates the outline of one possible solution, using the

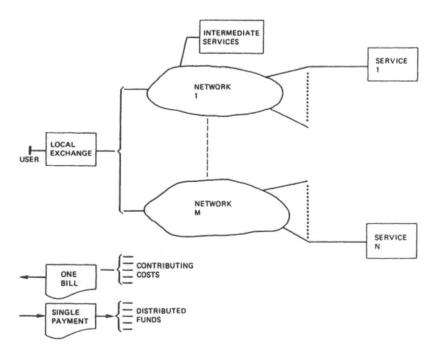

Figure 10.1 Integrated charging – the problem.

225

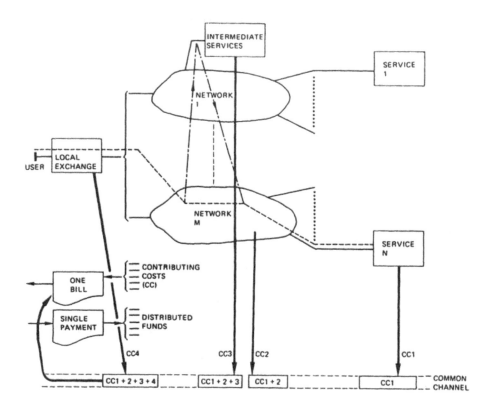

Figure 10.2   Integrated charging – a possible solution.

common channel to assemble an aggregated charge at the originating local network provider. The participating services and networks may, if they wish, record the transaction. In practice, they will probably record a selection, statistically based or based on "good payment" performance.

Payment performance aspects will be exacerbated by the legal difficulties involved, for example in calls via diverse networks in diverse countries to "foreign" services all controlled by different legal regimes: calls, moreover, that may be rerouted owing to fault and overload conditions.

It could be that the majority of subscribers will accept a consolidated bill with little or no justification, but it would be possible to provide a different but more costly service that allowed detailed account justification on request.

10.2 Choice and integration – security

In about 1961 the Kingston-upon-Hull Corporation first offered to its subscribers a "dial-a-disc" service. Naturally there seemed to be no need to bar access to dial-a-disc from incoming calls from the national network. This was in spite of the fact that, under the interconnection arrangements existing at that time, Hull was not reimbursed by BT (then the British Post Office) for BT calls completed in Hull. Shortly after dial-a-disc was introduced, the Post Office found it necessary to discipline some operators who were found to be persistently listening to Hull's dial-a-disc.

This is merely one illustration of the dangers to the owner of unrestricted automatic access to diverse, expensive services. Any increase in the variety of choice increases the dangers. All measures designed to prevent unauthorized use of the diverse services lessen the value of the services, by making them less user friendly.

We are, fortunately, already learning the lessons and developing the technology that enables this dilemma to be solved with the minimum of damage to the ease-of-use criteria of the ISDN. Before discussing the solutions, however, it is necessary to take the discussion of the problem a little further. Choice of service is not only a subscriber function. Security considerations demand that the service should be able to accept only authorized applications.

What is needed, therefore, is a series of measures that guarantee adequate (and, bearing in mind the overheads of their provision, only sufficiently adequate) security of access to both the user and the service. At various levels, it is possible to make assumptions of an increased ability on the part of the user but at all levels an optimum mixture of user friendliness and security is sought.

It will help to list the available security measures, and these are shown in Table 10.1 together with their implications for the user, the user terminal, the access network(s) and the accessed services. Most are self-explanatory and require no comment, but some are expanded upon in the paragraphs that follow. The implications for the network and the service indicated in Table 10.1 are not always certain. Doubtful implications are indicated as "possible facility" in the table. All the signature techniques, for example, are most likely to be verified by the service, although the network could perform the verification as a service to the service. Such behaviour, apart from probably being less secure, has not been considered by the regulatory bodies and turns the argument about monopoly service provision on its head.

Clear and recall is a quite powerful technique already used in, for example, remote maintenance access to telephone exchanges. It means that the unauthorized access must occur at the authorized user terminal and therefore excludes all but the most determined (and criminal) hackers.

Encrypted question and answer requires the user to carry an active device. This could be a "smart card", a credit card, for example, endowed with intelli-

Table 10.1 Access protection techniques.

Access protection	Subscriber requirement User terminal requirement	Network requirement	Service requirement
Network access code, service number	Memorize	Directory	Directory
Machine-readable card	Transport Terminal facility	Recognize	Recognize
User identity – PIN number	Memorize	Recognize/correlate	Recognize/ correlate
Clear and recall	None	Facility	Possible facility
Encrypted question and answer	Transport of device Terminal facility	Possible facility	Facility
Voice signature	Memorize	Possible facility	Facility
Graphic signature	None Terminal facility	Possible facility	Facility
Fingerprint	None Terminal facility	Possible facility	Facility
Encryption	None Terminal facility	Possible facility	Facility

gence. On sign-on, the network or service transmits to the device an encrypted message that the device decodes using a code word recorded during the previous application. The device then sends the required, encrypted response and is accepted by the network or service, which, finally, sends an updated code word for storage in the device and use during the next application.

Signature techniques can be made as selective as the unique personal attribute upon which they depend. To ensure this degree of selectivity involves devoting considerable bandwidth and machine complexity to the process.

Encryption is a final technique well known and always open to eventual penetration if the stakes are sufficiently high. Encryption can be no substitute for the preceding authorized sign-on procedures.

10.3 Integrating the customer

In Chapter 1 we presented the ISDN as a set of facilities very desirable to the user. Not least among the desirable attributes was the absence of wires; integrated access via a single route to diverse networks and services does much to tidy and simplify the home and the workplace. In subsequent chapters we have lessened the impact of this desirability to the user by indicating the many features of the ISDN that are already made available to the user by other means. It only requires a small bias against the ISDN to claim that the only remaining desirable feature is the almost complete absence of wires across the desk. Certainly the pre-existence of so many available networks and services makes the users' decision to choose the ISDN an evenly balanced choice. There is indeed an inherent economic advantage in that the single ISDN access replaces the many accesses to individual networks. The cost of these is much more than just the cost of wires. Then, too, there is the quality of basic access that combines the facilities of a small LAN with those of a small PABX.

An obvious way to make the ISDN desirable to the user is through pricing policy, and this will constitute one area for discussion. A second possible method is to convince the customer of the utility of ISDN despite the pre-existence of many of the services being offered by the ISDN. This will form the second subject area.

10.3.1 Pricing for integration

Like all the less technical areas discussed so far, pricing policy is inextricably linked with the other movements reshaping the industry coincidentally with the appearance of the ISDN. American deregulation has led to the "unbundling" of charges to avoid cross-subsidies. Much the same process is occurring in the UK as BT is required to show that it is not indulging in unfair competition by subsidising competitive service from other, more profitable areas of activity.

The methods of pricing include the following:
(a) Subscription for service: the equivalent of a club membership. By sub-scribing, the user acquires the right to benefit from the services offered.
(b) Connection charge: apart from a once and for all charge for providing the connection, the user may be required to pay a recurring charge to "rent" the "per user" equipment and services.

In most cases subscription and connection charges between them reimburse the network provider for the cost of providing the possibility of connection service to the user.
(c) Usage charges. A charge proportional to the amount of use made of the network services. This can be computed in one, or a combination, of:
  • time,[1] duration and distance, the traditional voice telephony method;
  • content, typically charging per bit;
  • or capacity, possibly a variation of either of the above methods in which the time, duration and distance or bit rate charge varies according to the capacity (bandwidth) used by the communication.

In devising a pricing policy for the ISDN the administration must take into consideration all of the following:
(a) Adequate return on investment.
(b) Price attractiveness: the degree to which the price must encourage the user to "join" the ISDN. The value the administration places upon ISDN

---

1. The term time and duration is not tautologous, the "time" being the time of day at which the communication takes place.

membership is an important constituent of this.

(c) The change in price necessary in competing services. For example, virtual circuit private networks offered via the ISDN compete directly with existing private circuit networks. The administration may wish to increase private circuit prices to encourage movement to the ISDN.

(d) Regulations. In adopting this strategy of price differentials the administrations must consider how to satisfy the regulators.

(e) Competing ISDN service offerings.

(f) Price of ISDN services offered to competitors. One aspect of ISDN attractiveness to the subscriber is the degree of accessibility to other ISDN users on other competing networks.

(g) The pricing strategy over time. Early users, offered limited access to few other users, experience a less valuable service than later users joining when the ISDN is subscribed to by large numbers of users. The value of the network to the user (and therefore perhaps its price) increases in proportion to the number of users connected.

(h) Pricing strategy (and regulatory constraints) for "ordinary" telephone users who may eventually use ISDN access but not ISDN features.

Among these considerations a fairly delicate set of choices has to be made to arrive at a pricing policy and strategy that will be one of the principal means of encouraging customer acceptance of the ISDN. In areas where the ISDN services are already available by conventional means the need for attractive pricing becomes particularly relevant.

Table 10.2 shows current European charges for connection and rental of ISDN access and illustrates that these lessons have not yet been learned.

Table 10.2 Connection and rental charges for basic and primary ISDN access[141].

Country	Connection charge ($)		Monthly rental ($)	
	Basic access	Primary access	Basic access	Primary access
Austria	140	1156	35	350
Belgium	176	12608	31	471
Denmark	239	2390	25	251
France	122	756	36	562
Germany	80	123	36	307
Ireland	627	6269	52	522
Italy	248	372	62	428
Luxembourg	224	2017	30	359
Netherlands	330	4396	38	379
Norway	213	2880	31	289
Portugal	191	824	27	824
Spain	282	8427	47	766
Sweden	379	6221	39	518
Switzerland	146	293	37	366
UK (BT)	613	4548	43	532

10.3.2 Utility to the customer, Centrex and VPNs

Consider, for a moment, the plight of the operating company salesperson discussing with the communications manager of a medium-sized company the possible merits to the company of ISDN service. Suppose the company has some five hundred employees, staff and clerical, located in three main sites and five area offices. The three sites each have computer facilities interconnected by a private data network. All eight sites have PABXs interconnected via a private voice network and the company subscribes to a viewdata service relevant to the industry. A closed user group of the viewdata service is specific to this company.

Our salesperson can offer ISDN connection to the company with probable (small) savings in the rental of private network services; can point to a significant simplification in the house wiring and desk wiring at all sites; and can offer a wider connectivity for data services such as mainframe-to-mainframe access for downloading such things as salary information to an external salary payment function. In general, however, the salesperson can offer only marginal improvements on the sophisticated features already enjoyed by the company. Indeed, in the early days of ISDN, the range of features offered by the ISDN may actually be less than those available by other means.

Picture now a different situation. Suppose that over the previous ten years the company has changed out its PABXs at two of the three main sites and three of the five area offices. In each case the company determined that, instead of replacing the PABX, it would use Centrex service. Centrex removed the need to house, power and maintain PABX equipment in the office. It carried slight advantages in cost of exchange line and private circuit rental. Centrex also proved more reliable as the public exchange reliability is, by design and by regulation, much greater than that of PABX equipment.

Over the years the emerging ISDN has presented the company with enhanced features, piecemeal, via the (ISDN compatible, presumably) Centrex services. There is, at this time, no need for the salesperson's visit as the company has been gradually acclimatized to the ISDN environment and is, by now, totally committed. These scenarios are illustrated in Figure 10.3.

Figure 10.3b shows the results of our salesperson's efforts should they be successful. The company has been enabled to combine its voice and data private networks at the expense of equipping all sites with an ISDN-compatible PABX. There are potential problems inherent in switching voice and data traffic through the PABX unless it has been specifically designed with this purpose in view. The company's private network, now provided via the ISDN, could consist of private digital circuits but may more probably be a VPN. Such a network consists of circuits, or packet-switched virtual circuits, provided by guarantee as required but used for other public network purposes when not required by the company.

By contrast, the Centrex-equipped company (Fig. 10.3c) begins to enjoy

(a)

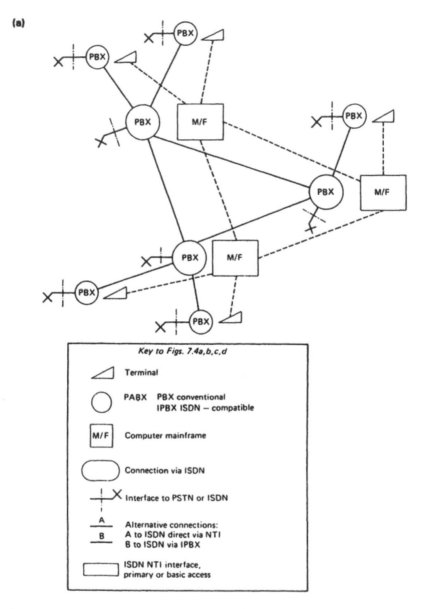

Figure 10.3 ISDN salesperson's problems. (a) Non-Centrex customer pre-ISDN. (b) Non-Centrex customer's ISDN objective. (c) Centrex customer pre-ISDN. (d) Centrex customer with ISDN.

(b)

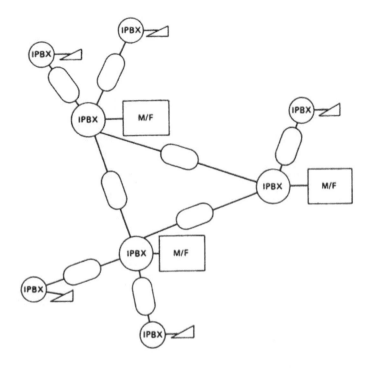

(c)

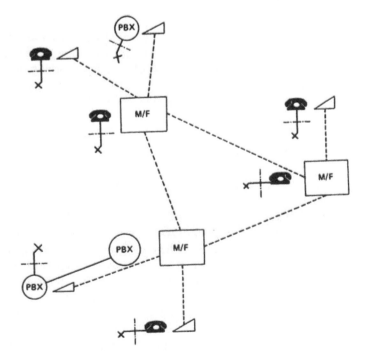

233

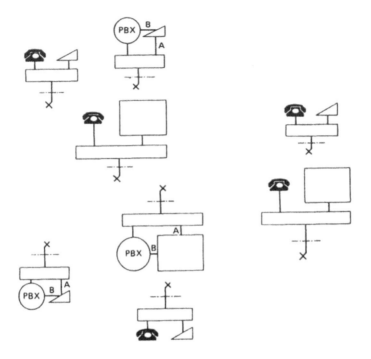

Figure 10.3 ISDN salesperson's problems. (d) Centrex customer with ISDN.

the ISDN as it permeates the public network, eventually approaching the situation shown in Figure 10.3d. There is no question of choosing a suitable ISDN PABX, the combination of voice and data traffic being dealt with by the public ISDN exchange, designed for the purpose. Figure 10.3d makes no attempt to illustrate the rich variety of functions available to the ISDN extensions.

10.3.3 Centrex and exchange line rentals

This discussion of the benefits of Centrex leads back to a further discussion of tariffs.

The purpose of this section is to provide an overall "feel" for the technical economy of Centrex. The conclusions drawn may not bear any direct relationship to existing tariff structures either in Britain or abroad but will give some indication of the direction in which tariffs might move over a period of years.

Figure 10.4 shows five different methods of telephone provision. No attempt has been made to illustrate the addition of data facilities (communicating PCs, facsimile, teletex, etc.) although these would be much simplified using the ISDN solutions of Figure 10.4d and e.

Figure 10.4a shows an extension telephone connected via an analogue

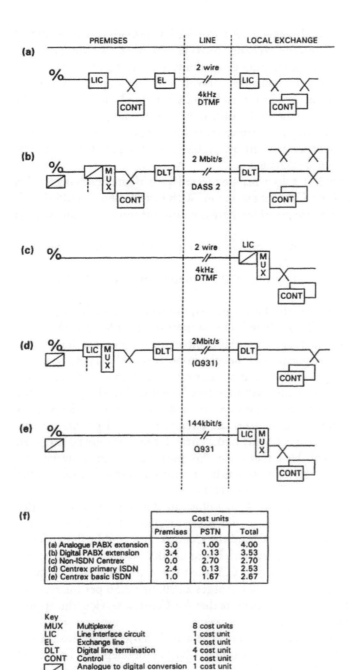

	Cost units		
	Premises	PSTN	Total
(a) Analogue PABX extension	3.0	1.00	4.00
(b) Digital PABX extension	3.4	0.13	3.53
(c) Non-ISDN Centrex	0.0	2.70	2.70
(d) Centrex primary ISDN	2.4	0.13	2.53
(e) Centrex basic ISDN	1.0	1.67	2.67

Key
MUX   Multiplexer   8 cost units
LIC   Line interface circuit   1 cost unit
EL   Exchange line   1 cost unit
DLT   Digital line termination   4 cost unit
CONT   Control   1 cost unit
▨   Analogue to digital conversion   1 cost unit

Figure 10.4 The technical cost of Centrex provision. (a) Analogue PABX and local exchange. (b) Digital PABX and local exchange. (c) Ordinary or Centrex Sub on digital exchange. (d) Centrex via primary rate interface. (e) Centrex via basic rate interface. (f) Technical economy comparison.

PABX to an analogue local exchange. A two-wire line provides a bandwidth of 4 kHz. Both the PABX and the exchange provide switching, facilities and control, although the exchange facilities provision is not used by the PABX extension. Note that the public exchange control is always duplicated (or the equivalent of duplication used by the exchange to give public network reliability), although to duplicate the PABX control the user would pay extra.

Figure 10.4b shows the same arrangement for a digital PABX connected to a digital exchange. In an ISDN environment DSS 1 (or its equivalent) would be used for CCS to the local exchange.

Figure 10.4c shows a subscriber connected to a digital exchange. This would be identical for a Centrex extension. With just pushbutton MF signalling only PABX facilities are possible; there is no opportunity for simultaneous voice and data.

Figure 10.4d shows Centrex provision using an ISDN primary rate interface.

Figure 10.4e shows Centrex (or DEL) provision over an ISDN basic rate interface.

A rough assessment of the hardware requirements for each solution is shown in Figure 10.4f. This is based on cost units per telephone as shown in the key to the figure. Excluded from this comparison are the cost of software and the cost of the analogue part of the telephone. It is probable that Centrex software will be subject to change at a faster rate than public exchange software and the network provider will therefore be entitled to charge more for Centrex provision. Further, the network provider will wish to deter users from transferring from dedicated private circuits to virtual private circuits purely on a basis of the cost economy of the latter. It can therefore be expected that Centrex tariffs will be inflated for this purpose. It can, however, be argued that the cost of the network has been reduced overall because of the less random nature of the traffic over VPNs.

With all these caveats Figure 10.4 does illustrate that the network provider could charge two to three times as much *per extension* for Centrex as for exchange line provision to a PABX and the Centrex user could still expect to effect an overall saving compared with use as a PABX. Thus, if the quarterly rental of an exchange line was, say, £25, then for a 200-line PABX with 20 exchange lines the network provider charges £500, or £2.50 per quarter per extension. The argument above suggests that for Centrex service, the appropriate charge might be £7.50.

Immediately the dilemma is apparent. The actual charge per extension will not be less than the £25 exchange line rental. Charges for telephone line rental are commonly set at a level that does not represent the value of the service. The network provider expects to get a return on call charges that compensates for the uneconomic rental. In this situation the evident technical economics of digital Centrex is obscured by the tariff policies that set the exchange line rental at too low a level.[2] Regulation is helping here. In the UK and other liberalized networks the regulator (OFTEL in the UK) is insisting that the network providers "unbundle" their charges, ceasing the subsidy of one service by another. Under this regime exchange line rentals are gradually rising to more economic levels, but only slowly as the regulator also is constrained by the negative political impact of too sudden a rise in line rentals.

The network providers may well take issue with this analysis, and with good reason, as it neglects the value of the enhanced services available with Centrex. It is nevertheless illustrative of the difficulties of marketing Centrex in a network in which the relation between price and costs is distorted.

10.3.4 The teleworker

Figure 10.5 illustrates a progression of arrangements for providing the capacity for company employees to work from home. The left-hand pair are factual and applications are known that have developed in this manner.

The first attempt involved a private data line from the employee's home to a PAD at the nearest PSS exchange. Transmission was at 1.2 kbit/s. Each end of the communication, the PC and the mainframe or the front-end processor, had to be equipped with asynchronous to synchronous conversion, error control and security provision. In using mainframe systems that were menu driven, the speed of response was such that users very soon became dissatisfied. This

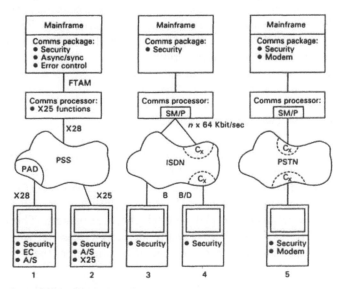

Figure 10.5 The teleworker.

---

2. Although rental charges differ throughout the world, what is described here is a universal problem. Few, if any, administrations charge a realistic amount for line rental.

was caused not only by the slow speed over the local X.28[142] line but by the excessive error control because of noise and interference.

The availability of X.25[15] cards compatible with IBM PCs allowed improvement by extending X.25 to the homeworker's PC. This removed the need for specific error control in the PC as the much more satisfactory error control of X.25 could be used throughout the connection. Faster speed (up to 9.6kbit/s) and better error control gave dramatic improvement to the services perceived by the user.

The next two examples use an ISDN channel, although this is (as shall be seen later) not essential to the argument. Now there is provision for synchronous communications from the PC over a 64kbit/s B channel if required. The error control (as with X.25) is inherent and the only additional requirement is the security provision. Combination of voice and data allows additional possibilities for this provision of security.

The difference in the latter of this pair of examples is that, by using Centrex and VPN, the homeworker has been integrated back into the company's private network and enjoys the use of the private network numbering scheme and all the private network features. The company sees all the homeworker's activities on the home installation.

The final diagram illustrates this first Centrex solution as it might be provided in the interim period before the deployment of the ISDN. The PC, the mainframe and all other communicating PCs in the network are equipped with compatible modems and, possibly, voice/data switching is possible at the terminal. These arrangements apart, the combination of Centrex and a digital network, with perhaps DSS 1-like signalling to the PSTN line termination, allows all the advantages of the previous solution.

10.4 "Normal service will be maintained"

In considering the ISDN it is necessary to be continually reminded that there will always be a significant number (for decades yet probably a majority) of subscribers whose only interest is in the "plain old telephone service (POTS)". In the early years of ISDN implementation, these do not present a problem, they remain subscribers connected to plain old telephone exchanges. There will, however, come a time, if the ISDN is even moderately successful, when the "reactionary subscriber" will be more conveniently connected to the ISDN part of the network.

Figure 10.6 depicts the options available. Figure 10.6a is the obvious solution and undoubtedly the most common in early implementations. Digital exchanges retain conventional line groups with digital-to-analogue conversion in the line circuit. The reactionary subscriber retains the existing telephone and is oblivious to the ISDN option. The network provider must

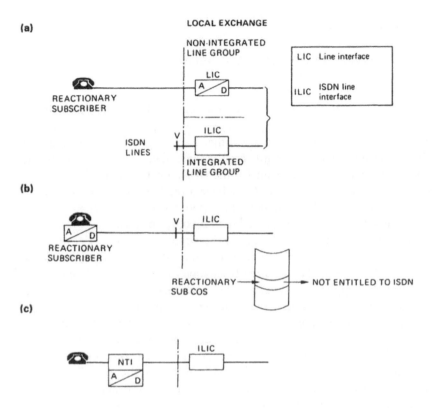

Figure 10.6 ISDN service to the non-ISDN subscriber. (a) Separate line circuit groups. (b) Simple ISDN telephone non-ISDN class of service. (c) Plain old telephone ISDN interface.

provide conventional analogue local links to the subscriber.

To obtain the local network savings of the ISDN the network provider must move the digital-to-analogue conversion to the telephone. In Figure 10.6b the subscriber is given a new telephone but is unable to use the ISDN features because the chosen class of service denies them. This may be a good option from a marketing point of view. If, however, the subscriber is to remain oblivious to the ISDN opportunities then Figure 10.6c is suitable, although the network provider has to obtain the subscriber's agreement to locating the NT1 equipment on the subscriber's premises. There may be a real problem here for all subscribers, ISDN or otherwise. Is it acceptable that the network provider uses electrical power drawn from the subscriber's mains supply? In the UK this has been the practice, without protest, for many telephone facilities, notably the Plan 7 secretarial sets. As explained in Chapter 3, this problem does not arise if the NT1 is powered from the exchange.

239

10.5 ISDN services

10.5.1 Services helping user and network: ISDN supplementary services

Supplementary services are services additional to ordinary telephony that the network or service provider may wish to provide, possibly for an additional charge. Many such services become rather easy to provide when CCS allied with digital transmission is extended to the subscriber's terminal. This was the proudest and most exciting claim for the ISDN. That some of these services, existing in the network without ISDN, have already been discussed indicates one of the problems of introducing the ISDN: it provides services that users have already been given or provided with by other means.

Table 10.3 indicates the richness of the supplementary services that will become available as the ISDN is introduced. Against each service is the identity of the ITU-T recommendations already existing or in process of being drafted to describe the supplementary service. For each service there can be three recommendations, the definition being structured as follows[84]: the stage 1 description describes the service in non-technical terms as it might be described to a potential user; the stage 2 description describes the service in terms of the messages involved and the actions to be taken; and the stage 3 description describes the service in the context of the signalling system involved in provision of the service. This is the full technical protocol for the service within that signalling system.

It is the scale of the opportunities with ISDN supplementary services that is so impressive. We have become accustomed to transferring our telephones around the office as we go to meetings or visit colleagues. The ISDN call transfer services permit calls to peripatetic executives to follow them around the world (CT, CFB, etc.). We need never again wonder who was calling as we fumble with our key in the front door: the identity of the caller will be on the little screen on our ISDN telephone or terminal (CLIP, possibly CW).[3] When we take our portable computer away with us and plug it in in a different country, the ISDN will recognize that we are at a new access of the worldwide network (TP).

That there are problems is undoubted. All the features dependent on the calling line identity are open to question because of legal objections to disclosure of this information (CLIP, CLIR, etc.). All these services have to have in-built facilities to prevent identification information being passed on from one public network to another. But some of the facilities dependent on these, such as ordering and paying for goods by telephone, depend on validation of the calling number plus, perhaps, obtaining a credit rating from a central database.

---

3. BT introduced call return as a free service during 1994–95 and it is proving to be a remarkably popular facility with the public, who are no doubt recompensing BT through increased telephone use.

Table 10.3 ISDN Supplementary Services[e.g. 71, 143–150].

Abbrev.	Title	Stage 1 ITU-T rec.	Stage 2 ITU-T rec.	Stage 3 ITU-T rec. for #7	for DSS 1
	Definition of supplementary services	I.250, 1988	–	–	–
	Introduction to stage 2 service descriptions for supplementary services	–	Q.80 1988	Q.730 1993	Q.950 1993
	Number identification supplementary services	I.251, 1988	Q.81	Q.731	Q.951
DDI	Direct dialling-in	I.251.1	Q.81.1	Q.731.1	Q.951.1
MSN	Multiple subscriber number	I.215.2	Q.81.2	–	Q.951.2
CLIP	Calling line identification presentation	I.251.3	Q.81.3	Q.731.3	Q.951.3
CLIR	Calling line identification restriction	I.251.4	Q.81.3	Q.731.4	Q.951.4
COLP	Connected line identification presentation	I.251.5	Q.81.5	Q.731.5	Q.951.5
COLR	Connected line identification restriction	I.251.6	Q.81.5	Q.731.6	Q.951.6
MCI	Malicious call identification	I.251.7	–	–	–
SUB	Subaddressing	I.251.8	Q.81.8	Q.731.8	Q.951.8
	Call offering supplementary services	I.252, 1988	Q.82	Q.732	Q.952
CT	Call transfer	I.252.1	–	–	–
CFB	Call forwarding busy	I.252.2	Q.82.2	Q.732.2	–
CFNR	Call forwarding no reply	I.252.3	Q.82.2	Q.732.3	–
CFU	Call forwarding unconditional	I.252.4	Q.82.2	Q.732.4	–
CD	Call deflection	I.252.5	Q.82.3	Q.732.5	–
LH	Line hunting	I.252.6	Q.82.4	–	–
	Call completion supplementary services	I.253	Q.83	Q.733	Q.953
CW	Call waiting	I.253.1	Q.83.1	Q.733.1	Q.953.1
CH	Call hold	I.253.2	Q.83.2	Q.733.2	Q.953.2
CCBS	Completion of calls to busy subscribers	I.253.3	Q.83.3	–	–
TP	Terminal portability	–	Q.83.4	Q.733.4	–
	Multiparty supplementary services	I.254	Q.84	Q.734	Q.954
CONF	Conference calling	I.254.1	Q.84.1	Q.734.1	Q.954.1
3PTY	Three-party service	I.254.2	Q.84.2	Q.734.2	Q.954.2
	Community of interest supplementary services	I.255	Q.85	Q.735	Q.955
CUG	Closed user group	I.255.1	Q.85.1	Q.735.1	Q.955.1
PNP	Private numbering plan	I.255.2	–	–	–
MLPP	Multilevel precedence and pre-emption	I.255.3	Q.85.3	Q.735.3	Q.955.3
	Priority service	I.255.4	–	–	–
OCB	Outgoing call barring	I.255.5	–	–	–
	Charging supplementary services	I.256	Q.86	–	–
CRED	Credit card calling	I.256.1	Q.86.1	–	–
AOC	Advice of charge	I.256.2	Q.86.2	–	–
AOC-S	Advice of charge: charging information at call set-up time	I.256.2a	Q.86.2	–	–
AOC-D	Advice of charge: charging information during the call	I.256.2b	Q.86.2	–	–
AOC-E	Advice of charge: charging information at the end of the call	I.256.2c	Q.86.2	–	–
REV	Reverse charging	I.256.3	Q.86.3	–	–
	Additional information transfer supplementary services	I.257	Q.87	Q.737	Q.957
UUS	User-to-user signalling	I.257.1	Q.87.1	Q.737.1	Q.957.1

Many of the problems and much additional complication within the supplementary service protocols are associated with the requirement that each use of the supplementary service can attract a charge. There is, indeed, a strong argument for criticizing administrations for their failure to offer ISDN access and ISDN supplementary services as a single package with little or no additional charges for the use of the individual services. Such a procedure would be much cheaper to implement and would make the ISDN a much more attractive proposition to the user. The network provider could still expect to obtain value for money invested through the increased call revenues. There is always a reluctance, as occurred in the UK with the failure of Prestel (see §10.1), for users to sign up for services whose charging arrangements are difficult to understand or to supervise and control.

A further problem, not unconnected with charging, is the problem of interactions between the different supplementary services. Completion of calls to busy subscribers (CCBS) may be affected in its operation by the presence of call forwarding supplementary service at the outgoing end. And if the two services complement each other will it be the calling user or the called user, or both, who will pay for the service? Similarly under what circumstances is a member of a closed user group able to go outside the group when constructing a conference?

All these problems are being addressed by a European standard, ETS 300 195, on supplementary service interactions[151]. In Europe, the free availability of the ISDN, complete with supplementary services, is a major plank of the open network practice directives. All the recommendations listed in Table 10.2 have their European counterparts as ETSI standards.

It is immediately obvious from this listing that many of the facilities are already available to PABX or Centrex users and could be made available from the SPC local exchange. Many more could be made available using existing, non-ISDN technology although the economics of their provision is made much more attractive by the ISDN. The particular merit of the ISDN is the ability to conduct simultaneous or sequential voice and data "conversations" from the same network access.

Figures 10.1 and 10.2 illustrated, incidentally, the use of intermediate services that have potential for particular benefits to the user and to the network provider. One of these is already in use in America, the UK and elsewhere, namely the "800 service". This service allows "free" connection by the caller to the called party, who pays for the call. In the ISDN implementation, the intermediate service contacted via CCS validates the call (the called party may place limitations on the source and distance of the calls to be connected), provides code translation and can provide calling identity, number, name, perhaps address, credit rating, credit card numbers, etc.

In the ISDN environment such intermediate services can be envisaged for a variety of purposes. Protocol conversion between data terminals to systems that are not "open" could be such a service.

242

Note that Table 10.3 has two columns for the stage 3 protocols; there is a requirement for a different protocol for each CCS signalling system. As noted elsewhere, what a pity.

10.5.2 Services helping the network provider

The features of the ISDN that are of interest to the user are, of course, of utility to the network provider. Just as the user can reconfigure a VPN, so can the administration, with suitable security provision, reconfigure the public network. Just as the user has access to data concomitantly with a voice call, so the administration has access to its network data. All the network maintenance, network management and customer service features treated at length in reference 2 are more easily available by using the ISDN. It is not intended here to deal with the network provider aspects in detail as reference 2 constitutes a reasonable introduction. Some details only, specific to the ISDN, will be highlighted.

Provision of SPC and CCS allows the man–machine interface considerable freedom to manipulate the network. Consider a subscriber requesting a change in service provision for example. With existing SPC or CCS features the service desk can act on a telephone request, introduce the new service and alter the subscriber's class of service and billing data. But there is no authorized record of the transaction and an administration may be, rightly, reluctant to change services without written authorization from the user. With SPC the service desk has CLI confirming the subscriber's identity and authorization status, but with ISDN it can ask the user to confirm by, say, obtaining a credit card number (passing a credit card through a reader perhaps), and can demand further authorization codes and passwords as is deemed appropriate. When the technology of graphics input is mature the ISDN will be ready to transmit signatures.

We have already indicated the benefits of ISDN to network management by combining the various network resources for voice and various forms of data into the one resource that may be managed in its entirety. Thus, circuits dedicated to circuit-switched data at night may send packet-switched data or voice by day. Network elements may be sometimes private network entities, sometimes public network resources.

10.6 The information superhighway

For Christmas 1994 the author's family gave him subscription to the Internet. Unlike the ISDN the Internet has caught the imagination of the press and, over the past three years, has received tremendous attention in the media and achieved enviable success in the market place. The service is very similar to viewdata but is based on communication between personal computers using ordinary ASCII-coded text for constructing text and images rather than communication between machines with visualization by means of images built up from pixels more suited to the television screen.

The Internet subscriber joins a server organization, which issues a unique user address. All communications to and from the user are via a service access point reached by dialling over the PSTN. Thus, like viewdata, incoming messages are only received when the subscriber connects to the service.

The popularity of the Internet has been assured by the rich variety of information available from it. Most of this information is available free, and much of it is of an interactive nature. Gradually, information providers are seeking to recover access fees, but there is a mass of information available free or relying on use of further services to recover the cost. For example, it is now possible to buy and sell shares in the US and in the UK at least, over the Internet. The cost of providing the dealing data is recovered from the commission on the sale. There are lessons here for the network providers who insist on charging for supplementary services on the ISDN.

At present, users of the Internet are equipped with analogue modems with speeds up to 19.2 kHz, so that downloading, say, ITU-T Recommendation Q.931[31] will take a little while.[4]

Of course, the Internet subscriber will be well advised to subscribe to the ISDN when the speed of access will be much more user friendly. At present, however, as we have seen, the tariffs sought by the network providers preclude this. Here too is an instance where call tariffs based on time and distance will be unsuitable: they ought, with justice, be based upon the quantity of data.

10.7 The mobile network and the ISDN

10.7.1 The development of the cellular mobile system

It is surprising that, having mentioned mobile radio systems in Chapter 1, there has been only one further reference to the subject, in Chapter 8, until now. It is necessary to discuss, however, how the massive introduction of mobile networks affects and interrelates with the ISDN. To do so I will start with a little history of the emergence of mobile networks that has occurred at substantially the same time as the ISDN technology has been developed.

The first edition was only able to comment on the first two years' experi-

---

4. ITU-T recommendations were, until recently, available from several Internet addresses. The ITU now seems to have closed off this free method of obtaining their documents.

ence of cellular radio in the UK and noted that the popularity of the service had exceeded all expectations. There was even a time when the upwardly mobile would carry a wooden model of a mobile phone when demand was so great that they could not obtain a real one.

Cellular radio is a development that has surprised everyone by its popularity. A local development intended to overcome a local difficulty has provided the greatest success story since perhaps the invention of the telephone. In the large stretches of thinly populated northern Scandinavia there have always been real problems with keeping in touch. Communications in these regions can be a matter of life and death. During the 1970s the Nordic countries co-operated in the development of a radio telephone system that would provide communications throughout Scandinavia without the need for an increased investment in the terrestrial network and without excessive use of the precious resource of scarce radiofrequency spectrum. The result was a cellular system based on the use of relatively low-powered base stations transmitting to and receiving from mobile user stations, usually provided as car telephones. The frequencies used in one cell could be used again in any but adjoining cells. The principle is illustrated in Figure 8.1. To everyone's surprise the system became most popular in urban areas for use by business people and there had to be almost immediate upgrading of the system to permit much smaller and far more numerous cells to provide adequate coverage in urban areas. Very soon the system, or one like it, was in use in the UK, in other European countries and in the USA. In the UK, the system chosen for early implementation was total access communication system (TACS), which, like the original Nordic mobile telephone (NMT), was an analogue system.

A cellular radio phone is not much bigger than a paging receiver but far more complex, allowing ordinary local, trunk and international telephone calls to be originated as well as received. One of the difficulties with all radio systems is the hard fact that frequencies are limited. Good radio coverage of a wide geographic area at the very high frequencies used for these services can only be obtained by siting the main transmitters and receivers at as high a point as possible – on top of the tallest local building or on a nearby mountaintop. This is fine so far as it goes, but if there are available only enough frequencies to allow, say, 30 conversations at a time, whereas it is necessary to be able to carry over 100 calls at a time, then the mountaintop site with its wide area coverage is no longer quite so suitable. It is necessary to be able to divide up the geographic area so that it may be served from more than one main station. Antennae may, for example, be made more directional, so that the one main site acts as if it were several different main stations, each covering say 90° of arc. By reducing the power of main stations, and in some circumstances placing their antennae only half-way up a building instead of at the top, it is possible to reuse frequencies in other areas not very far from where they are already in use. This means that the available frequencies, usually the resource that runs out first before all traffic requirements can be met, can look after far more simultaneous conversations than before.

This repeated reuse of frequencies is one of the main differences between cellular systems and their predecessors. The other is that mobiles on a cellular system are able to change their own operating frequencies as they move, a facility called "roaming". As the mobile moves around a city the strength of the radio signal varies at the receiving station carrying the call. Whenever the signal strength gets too low for satisfactory operation all neighbouring fixed stations are instructed automatically to listen out for the signals. If one of the neighbouring cell area stations receives the signal better than the original "home" station, control is passed over straight away to this other station, the mobile set changes its transmitting and receiving frequencies to ones used by the new master and the network control station switches all the external circuits over so that whoever the mobile was talking to, either a subscriber on an ordinary telephone system anywhere in the world or another mobile, is now connected via this new controlling station (Fig. 8.1).

Cellular operations began in the UK with two competing nationwide cellular radio systems, Cellnet and Vodafone, each with its own network of linked radio main stations and switching centres. But by a somewhat complex series of regulations the radio-phones themselves are provided and installed and "airtime" is paid for via service providers and their local contractors.

One of the much publicized difficulties experienced by users of cellular services has been brought about by the great popularity of these services. There are now far more mobile phones on the road in some areas than the planned traffic-handling capacity of the networks can reasonably support. The London orbital motorway, the M25, has been congested with traffic ever since it was opened and has been congested for cellular users for almost as long. Cells are, however, continually being subdivided and new frequencies being made available in order to increase traffic-handling capabilities and remove congestion, but at peak times and in some areas it is still difficult to make a satisfactory and uninterrupted call from a mobile.

It is, of course, also not possible to use a present-day UK car phone while driving anywhere in continental Europe and Continentals cannot use their car phones here in the UK either. However, a completely new digital cellular system that provides pan-European coverage has now been designed. Users are able to use their car phones wherever they are in Europe and will be billed for calls made. This new system, designed to replace all of Europe's existing analogue cellular mobile systems, is called by the name of the study group that drew up its specifications, Groupe Spécial Mobile (GSM).[5] The developers and the makers of GSM equipment are now marketing GSM worldwide and have renamed it the global system for mobiles.

In retrospect, it is, of course, a pity that the mobile systems, although they began in the digital era, were initially implemented using analogue techniques. There is now a need to support two systems: the existing analogue

5. GSM was the first ETSI project team and was working before ETSI was formed.

systems and GSM and other, newer, emerging digital systems. These successor systems, now in the process of standardization, are variously called fixed public land mobile telecommunications service (FPLMTS) by the ITU-T and universal mobile telecommunications service (UMTS) by ETSI. These systems are intended to support voice and data over mobile communication links capable of handling bandwidths up to 2 Mbit/s.

In truth, however, making the mobile system digital does not do all that much to make the mobile systems more compatible with the digital PSTN. Because of the scarcity of radiofrequencies, we cannot afford to allocate 64 kbit/s to a digital mobile telephone conversation. The digital encoding laws used for mobiles are therefore entirely different, succeeding in encoding voice in, for GSM, 13 kbit/s.

The progression to digital mobile radio in the USA has been rather different. GSM and the Japanese system personal digital cellular (PDC) use time-division multiple access (TDMA) using a different bearer frequency for each coded stream of time-multiplexed communication channels. The equivalent system in the US is advanced digital cellular (ADC), but there is a rival system in use in the US, code division multiple access (CDMA), in which each sender encrypts the signal with a code and the receiver decodes the received signal using the same code so that many more bit streams, each using a different encoding/decoding code, can be carried within the same frequency band. Figure 10.7 shows the principle at the receiving terminal.

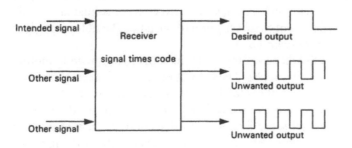

Figure 10.7 Principle of code division multiple access.

10.7.2 Cellular systems and the ISDN

In this account of cellular mobile systems two important aspects have been identified: digital cellular systems encode voice signals in a different way to that used in the fixed network and the mobile telecommunications switches have a new function – they have to support roaming.

The fact that the encoding laws are different introduces another codec function into the combined network. Because the exchange is different, a trend that emerged when studying Figure 7.2, i.e. for the broadband public switching network to develop in a way that diminishes, perhaps almost to vanishing point, the number of conventional real-time switching centres (telephone exchanges) in the network, is endangered. For the mobile network there is a need to retain the telephone exchange at least because of the need to provide roaming.

There is also a less obvious new requirement on the telephone exchange dealing with mobiles. This is an increased traffic capacity requirement. In conventional circuit switching, the exchange only has to work at call set-up and clear-down times, and its processing power, in terms of BHCA (busy hour call attempts), is determined by the expected incidence of calls. Only its size, in terms of lines and trunks and switch capacity, is influenced by traffic call-holding times. In the mobile exchange processing power also is influenced by call-holding times as the exchange has to work every time the mobile moves out of range of the current base station.

Chapter summary

In continuing to examine the problems inherent in the introduction of the ISDN this chapter was devoted to the more administrative, less technical problems. Chief among these are the problems of providing the user with a consolidated bill for the diverse variety of networks and services that may be involved in a single ISDN "call". The absence of such a consolidated bill will, it is claimed, act as a deterrent to subscription to the ISDN. Possible methods of solution based on CCS were outlined.

The wide variety of choice available over the ISDN brings with it another problem, that of authorizing access. The various methods of access protection available to the ISDN were outlined and, where necessary, discussed.

These two subjects were presented as likely disadvantages of the ISDN as perceived by the potential subscriber. The third section considered ways of encouraging ISDN subscription. One such means is an attractive pricing policy and the constraints on such a policy were discussed. More direct methods were then treated and, as indicated in Chapter 9, Centrex services were presented as an important method of gaining customer interest. Discussion of Centrex led back to the problems of tariffs, particularly in relation to the distortions introduced by a universal tendency to undercharge for exchange line rental. Partly in relation to Centrex, the provision of services to the person working from home was discussed. "Teleworking" is set to become a potent new method of employment and depends, to quite a considerable extent, on telecommunications. The ISDN could become essential to such people.

It was also necessary to consider the needs of the "reactionary" subscriber who does not want, and never will want, the ISDN. Eventually such subscribers will be served by the ISDN when fully deployed and their possible treatment was outlined.

The ISDN service offerings were looked at, picking out one or two for special comment, illustrating the wonderful variety of features available to user and network. This led to another wealth of offerings that have come independently of the ISDN via the Internet but which will be wonderfully enhanced when access to the Internet is available over the ISDN.

Perhaps belatedly, the discussion turned to mobile systems and, in particular, cellular mobile systems, leading to the conclusion that the development of these systems and their integration into the ISDN has profound effects on the possible development paths for the network.

What next?

The network architecture

In the final chapter of the first edition a concept for system design in the ISDN era based on a kind of planar network was introduced (Fig. 10.8). The idea was extended in the light of the development of SDH in the second edition of another book[2] by introducing the concept of a background network.

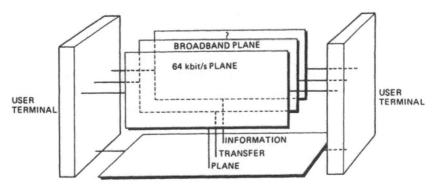

Figure 10.8 General arrangement of a proposed network.

The network could consist of two portions. The first would be a background network, dimensioned to carry the bulk of the traffic and to be fully occupied most of the time. This network would be semipermanently connected using DCCs as the switching nodes. The second, foreground, network would provide the remainder of the necessary network resource using traffic routes set up, as now, on a call-by-call basis. The arrangement is illustrated in Figure 10.9.

It is evident that this proposal is not new. Much of what has been said about

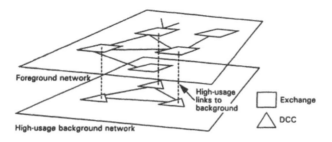

Figure 10.9   Principle of the background network.

the SDH indicates that the concept includes within it the assumption that all communications traffic is using the hierarchy. Figure 7.2, indeed, contains much of the same idea, suggesting that much of the traffic would be switched via cross-connects rather than by conventional telephone exchanges. Describing the background network is probably only formalizing concepts inherent in SDH. Again, emerging network offerings such as VPN are predicated upon the assumption that the network provider will use the knowledge of the existence of dedicated traffic directions within the customer's VPN. By this means the network provider will save on the overall network resource provided because, now, a significant part of the network traffic, the VPN part, is known and no longer random. To achieve this the network provider must be able to manage the VPN portion of the traffic on a semipermanent basis, and how better to do this than on a background network of DCC nodes.

That these ideas, now at least four years old, are still valid is illustrated by Figure 10.10 taken from a recent study[152]. This shows the continued validity of the concept of the information transfer plane (ITP) introduced in Figure 10.8 and of the idea of the background network carrying the larger proportion of the traffic that is less frequently in need of switching.

Meanwhile, what of the telephone exchange? The background network proposal would make it much smaller as it now only has to skim the busy hour traffic off the top of the bulk of the traffic that is destined for the background network. The migration of intelligence out into the local network has, arguably, made the exchange less intelligent. This will not be the case, however, until network providers can agree to use a compatible set of CCS protocols for both user and network. Compatibility of user and network protocols is addressed in ITU-T Recommendation Q 699[153]. In the interim the exchange will be the point of protocol conversion. Similarly, the introduction of the (limited) intelligent network concept has moved additional intelligence that might otherwise have been destined for the exchange to the service control point (SCP).

Tending against this trend is the integration of the cellular mobile services into the ISDN. These will still require their version of the local exchange

250

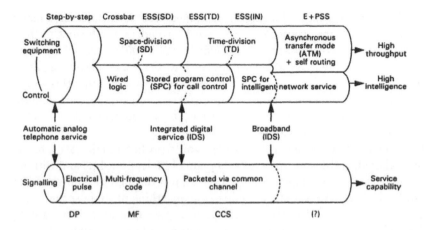

Figure 10.10 History and possible future of switching.

because of the need for the roaming facility. Although ISDN in the fixed network can envisage the exchange moving towards a centrally provided function, the needs of the mobiles may force its retention as a local exchange.

Nevertheless it is as well perhaps that the realignments in the industry and the massive cost of development of the present generation of digital SPC exchanges have delayed the development of their successors. The successor systems, when they are developed, will be smaller. They will probably be based on high-capacity ring structures in which each node will be equipped with limited intelligence for a limited objective and will rely on other nodes on the ring for services when greater intelligence is required of it.

Guiding development

Throughout, this book has included instances of political considerations increasingly impinging upon technical developments, largely because of the generally beneficial moves towards liberalization and privatization. An example of current interest is the licence restrictions on BT from participating in the cable TV market. At the 1995 party conference, the leader of the main opposition party announced that BT would install Internet facilities, free of charge, in every school and college in the land in return for the lifting of this restriction. After a four-year pause during which BT has done little work on fibre to the kerb technology[154], its interest, dampened by the licence restriction, is likely to be rekindled.

Why have we had to wait so long and why have not the network providers been more eager to sign us up for the ISDN? Much of the wait has been caused by the long process of producing the ISDN standards. It was during this production process that first the connectionless data interchange over the D channel and then the prospect of a single CCS system were lost. Both these losses are attributable to the attitude of the network providers, the former to their lack of enthusiasm to offer a feature for which they would not be able to exact a call charge and the latter to their reluctance to share a common signalling channel with the user because of the supposed security issues that this would entail. In addition, in much of Europe and America at least, there are a multiplicity of network providers whose rivalry has, in part, contributed to the long delay in the coming of the ISDN.

It is arguable that the network provider would profit from the ISDN, forcefully marketed, at the same rates as for a plain old telephone line and without additional charges for supplementary services, just through the generation of the additional calls that the ISDN would cause. Just think of all those messages, replacing a non-revenue-earning, ringing, no-reply signal to the calling user, each of which might generate (in the UK, is now generating) a return call when the called user returns to the telephone. Just think of the additional calls that would be made if users could travel with their personal number so that potential callers could still make contact without having to know a user's new number. This facility is available as a supplementary service (TP) that is extremely complex purely because of the requirement that the TP supplementary service should be charged at an extra rate.

The delay in bringing ISDN access to the user has been largely because of this issue of complexity. The ISDN standards represent a new record in ingenuity and complexity. They have not been easy to produce and agree even without any of the additional complexity insisted upon by network providers to provide security of signalling and means of additional charging. They are complex in their own right.

Architects and civil engineers (the better ones) have a deep appreciation of the excellence of simplicity of line and style. The aesthetics of design seems to be a concept not now considered by telecommunications or data processing designers. This used not to be the case. When every feature could only be realized by a number of relays, transistors or valves, the designer considered long and hard how to provide the feature with the least possible number of components. With large-scale integration and SPC, this economy in design has been discarded as an objective. Memory is cheap and components come ready packed in millions. What has not changed, however, is the ability of the human brain to understand complexity, and all these designs must still be understood by someone.

Methods have been devised to provide computer assistance to the designer of complex systems, but our designs are now reaching the limit where such aids are of any use and are certainly well past the limit where even a simple terminal equipment can be understood by a single engineer.

Perhaps it would have been better on the whole if the authors of ITU-T Recommendations Q.921[34] and Q.931[36], the ISDN OSI layers 2 and 3 protocols, had stopped when the documents exceeded 20 pages and looked around for a better way of doing this. These Recommendations are now the essence of simplicity compared with some of their successors prepared to describe mobile radio telecommunication services.

A look at the protocol stacks proposed for the B-ISDN also shows that things are not getting any better, or certainly not getting any simpler. New protocols are proposed to permit the flexible communication of speech as well as data by bandwidth-efficient methods (ATM). The network transmission is seen to be SDH, which, as we have seen, is an elegant means of transmitting large volumes of diverse information and extracting a single communication from the transmission without having to disassemble and reassemble the complete multiplex. Such developments are going a long way to fulfil the predictions in one of the authors' books[2], touched on in Chapter 5 (Fig. 5.7), that the network intelligence would migrate towards the terminal. Despite the difficulties presented by mobile communication discussed in section 10.7, the network could become an integrated whole depending in the main on relatively simple digital cross-connects and rather intelligent add-drop multiplexers.

Thus, we seem to be getting the architecture of the network right, but the detail, the application protocols, it could be argued, is far too complex. This appears to be partly because we are entrusting the design to too large an extent to international committees whose interests are not primarily the elegance of design of the network but the requirements of their sponsoring organizations.

The concept of CCS was the idea of one man. The idea of pulse code modulation was the idea of one man. Both of these were in days when an individual could understand in fair detail the working of a telephone exchange. It is arguable now that some of the failure to deliver usable user-friendly systems that are economic to implement is because of the drift into unnecessary complexity. We are asking the right people to design our networks but asking them to do so in the wrong forum and with the wrong priorities.

In two editions of the same book we have seen excellence in concept lead to disappointment in application because of a diversity of problems, far too many of which are not of a wholly technical nature. Our new technology was planned well but has turned out to be not what we imagined, expected or even desired. The final word, as in the first edition can again be left with Lewis Carroll:

For the Snark *was* a Boojum; you see.

The Hunting of the Snark, Fit the Eighth

Glossary

Not all the expressions explained in the Glossary are used in the book.

#6	ITU-T common channel signalling system no. 6
#7	ITU-T common channel signalling system no. 7
AAL-IDU	AAL interface data unit
AAL-SDU	AAL service data unit
ABR	available bit rate ATM service
ADC	advanced digital cellular
ADC	advise duration and change
AIS	alarm indication signal
AMI	alternate mark inversion
AMS	audio/visual multimedia services
ANSI	American National Standards Institution
APS	automatic protection switching
ARPANET	Defense Advanced Research Projects Agency Network, a wide area network that was intended to link research centres of the US Government Defense Department
ARQ	automatic repeat request
ASCII	United States of America standard code for information interchange
ASN	abstract syntax notation; ASN 1 is specified by ITU-T Recommendation X.208 as a semiformal tool to define protocols.
ATDM	asynchronous time-division multiplexing, early Bell Laboratories name for cell relay.
ATM	asynchronous transfer mode
AU	administrative unit of an SDH system
AUG	administrative unit group, an SDH term
B-ISDN	broadband ISDN
B8ZS	bipolar with eight-zero substitution, line code
Bellcore	Bell Communications Research
BHCA	busy hour call attempts
BIC	#7 bearer identification code
BIP	bit interleaved parity, SDH term; BIP-8 and BIP-NX24 are used
BOC	Bell Operating Companies
BSI	British Standards Institution
BT	British Telecom
BT	burst tolerance, an ATM service descriptor
CAC	Commercial Action Committee of CEPT
CAD/CAM	computer-aided design/computer-aided manufacture

CATV	common antenna TV
CBR	continuous bit rate ATM service
CCH	Harmonization Coordination Committee of CEPT
CCIT	Comité Consultatif International Radio (now ITU-R)
CCITT	Comité Consultatif International Télégraphique et Téléphonique (now ITU-T)
CCS	common channel signalling
CCTS	Coordination Committee for Satellite Telecommunications of CEPT
CDMA	code division multiple access
CDVT	cell delay variation tolerance, an ATM service descriptor
CEPT	Comité European Postes et Telecommunications
CIV	cell interarrival variation, a name for jitter used in ATM
CLP	cell loss priority field, of the ATM cell header
CLTA	Liaison Committee for Transatlantic Telecommunications of CEPT
CNET	Centre National d'Etudes des Telecommunications
CNM	customer network management
CPE	customer premises equipment
CRC	cyclic redundancy code and cyclic redundancy check
CS	convergence sublayer, ATM term
CSMA-CD	carrier sense multiple access with collision detection
CT1	first-generation cordless telephone systems
CT2	second-generation cordless telephone systems
DASS	digital access signalling system, the UK version of DSS 1
DBP	Deutsche Bundespost
DBS	direct broadcasting by satellite
DCC	data communications channel
DCC	digital cross-connect
DCE	data circuit-terminating equipment
DCS 1800	digital communication system at 1800 MHz
DDI	direct dialling in to PABXs
DECT	digital European cordless telecommunications system
DPNSS	digital private network signalling system, the UK equivalent of DSS 1 in the private network
DQDB	distributed queue dual bus
DSS 1	digital subscriber signalling system no. 1
DTE	data terminal equipment
DTI	Department of Trade and Industry
DUP	data user part
ECSA	Exchange Carriers Standards Association of the USA
EEC	European Economic Community, now called the EU
EFT	electronic funds transfer
EFTPOS	EFT at the point of sale
EIA	Electronic Industries Association
EPD	early packet discard, ATM congestion control technique
ERMES	European radio message system
ETSI	European Telecommunication Standards Institute
EU	European Union
FCC	Federal Communications Commission
FCS	frame check sequence
FDDI	fibre-distributed data interface
FDM	frequency-division multiplex
FPLMTS	fixed public land mobile telecommunications system, the ITU-T name for UMTS
FSK	frequency shift keying

FTAM file transfer, access and management
FTP file transfer protocol, a higher level application program associated with TCP-IP
GFC generic flow control
GSM Groupe Spécial Mobile, global system for mobile communications
HDB3 high density binary modulo 3
HDLC high-level data link control
HEC header error control octet of the ATM cell
HEC header error control field, of the ATM cell header
HRC hypothetical reference circuit
I frame information frame, LAPB information frame format
IDA integrated digital access, the early UK version of the ISDN.
ILMI Interim local management interface, an American ATM temporary expedient
IN intelligent network
ISDN UP ISDN user part
ISI intersymbol interference
ISO International Standards Organization
ITA 1 international teletypewriter alphabet no. 1
ITP information transfer plane
ITU-R International Telecommunication Union, Radiocommunication Standardization Sector
ITU-T International Telecommunication Union, Telecommunication Standardization Sector
Jitter short-term non-cumulative variations of the significant instants of a digital signal from their ideal positions in time
LAN local area network
LAP link access procedure
LAPB balanced link access procedure
LAPD balanced link access procedure in the D channel
LATA local access and transport area
LSI large-scale integration
MAN metropolitan area network
MCL Mercury Communications Ltd
MCR minimum cell rate, an ATM service descriptor
MFJ modified final judgement of the Computer II enquiry
MMC man–machine communications
Mobitex mobile data system (Sweden)
modem *modulator/dem*odulator
MPOA multiprotocol over ATM
MPT1327 Ministry of Posts and Telecommunications (UK) signalling standard
MSOH multiplex section overhead, SDH term
MSU #7 message signal unit
MTP #7 message transfer part
NCTE network carrier terminating equipment
NMT Nordic mobile telephone
NNI network node interface, the interface between switching nodes or between networks.
NSP #7 network service part
OFTEL Office of Telecommunications
OMAP #7 operation and maintenance application part
ONP open network provision
OSI open systems interconnection model of the ISO
PAMR public access mobile radio

PCM	pulse code modulation
PCN	personal communications network
PCR	peak cell rate, an ATM service descriptor
PDH	plesiochronous digital hierarchy
Pl OAM	loss of physical layer OAM cells in ATM
PMR	private mobile radio
POCSAG	Post Office Code Standardization Advisory Group
POTS	plain old telephone service
Primitive	the term used to describe a defined communication involved in the interaction between two adjacent layers in the OSI model. ITU-T (Recommendation X.210) defines it as: "Service primitive; primitive; An abstract, implementation independent interaction between a service-user and the service-provider." [161]
PSDN	public switched data network
PSS	packet-switched service
PSTN	public switched telephone network
PTI	payload type identifier
RACE	Research and technology development in Advanced Communications technology in Europe
RM	resource management, an ATM flow control message
RPI	retail price index
RR	receive ready, LAPB response
RSOH	regenerator section overhead, SDH term
S Frame	supervisory frame, LAPB supervisory frame format
SABME	set asynchronous balanced mode, LAPB command
SAPI	service access point identifier, DSS 1 term
SAR	segmentation and reassembly sublayer, ATM term
SCCP	#7 signalling connection control part
SCP	service control point
SCR	sustained cell rate, an ATM service descriptor
SDLC	synchronous data link control, IBM packet transfer protocol
SDU	service data unit, ATM term
SEAL	simple and efficient adaptation layer, ATM AAL type 5
SIF	#7 signalling information frame
SIO	#7 signalling information octet
SLS	#7 signalling link selection
SMDS	switched multimegabit data services
SN	service node, what used to be called the local exchange (LE)
SNA	systems network architecture, IBM rival to OSI
SNI	service node interface
SNR	signal-to-noise ratio
SOH	section overhead, SDH term
SONET	synchronous optical network
SPC	stored program control
STM	synchronous transfer module of an SDH system
STS	synchronous transport signal, SONET term equivalent to a STM in SDH but one-third of the size
STS-1	synchronous transport signal level 1 of a SONET system
TACS	total access communication system
TASI	time assignment speech interpolation
TC	transmission convergencer, ATM term
TCP	transaction control protocol
TCP-IP	transaction control protocol – internetwork protocol
TDMA	time-division multiple access

TEI   terminal end point identifier
TETRA   trans-European trunked radio system
TMR   trunked mobile radio
TRAC   Technical Recommendations Application Committee of CEPT
Tributary   SONET term equivalent to a TU in SDH
TSC   #7 time slot code
TU   tributary unit of an SDH system
TUP   #7 telephone user part
U Frame   unnumbered frame, LAPB unnumbered control functions frame format
UA   unnumbered acknowledgement, LAPB response
UBR   unspecified bit rate ATM service
UMTS   universal mobile telecommunication service, the ETSI name for third generation mobile radios
UNI   user–network interface
UPT   universal personal telecommunications
UTC   co-ordinated universal time
VADS   value-added data services
VANS   value-added network services
VBR   variable bit rate ATM service
VBR-NRT   variable bit rate – non-real-time ATM service
VBR-RT   variable bit rate – real-time ATM service
VC   virtual container of an SDH system
VCI   virtual circuit identifier, ATM cell field
VDU   visual display unit
VF   voice frequency
VISYON   variable intelligent synchronous optical network, a German network development
VPI   virtual path identifier, ATM cell field
VPN   virtual private network
VT   virtual tributary, SONET term equivalent to a VC in SDH
WAN   wide area network
Wander   long-term non-cumulative variations of the significant instants of a digital signal from their ideal positions in time
WATTC   World Administrative Telegraph and Telephone Conference
ZBTSI   zero byte time slot interchange

References

1. Orbell, A. G. 1981. Digital transmission in the local network. Paper presented at the International Conference on Telecommunications Transmission, London, March.
2. Ronayne, J. 1991. *Introduction to digital communications switching*, 2nd edn. Pitman and UCL Press.
3. International Standards Organization 1982. ISO/TC97: Information processing systems, open systems interconnection – basic reference model, draft international standard ISO/DIS 7498, April.
4. ITU-T Recommendations Q.310 to Q.332: signalling system R1.
5. ITU-T Recommendations Q.400 to Q.490: signalling system R2.
6. ITU-T Recommendations Q.251 to Q.300: signalling system no. 6.
7. ITU-T Recommendations Q.700 to Q.821: signalling system no. 7.
8. ITU-T Recommendations Q.120 to Q.180: signalling systems no. 4 and 5.
9. Povey, P. J. 1979. *The telephone and the exchange*. Pitman.
10. Flowers, T. H. 1976. *Introduction to exchange systems*. John Wiley.
11. ITU-T Recommendations Q.850 to Q.1063: digital subscriber signalling system no. 1.
12. Ericsson, undated. Telerica: a series of telecommunications case studies, Book no. LZT 10935ue.
13. Daniels, E. E. 1981. Telex switching and signalling, *Post Office Electrical Engineers Journal* 74, Part 3.
14. ITU-T Recommendation X.75: Packet-switched switching system between public networks providing data transmission services, Revision 1, 1993.
15. ITU-T Recommendation X.25: Interface between data terminal equipment (DTE) and data circuit-terminating equipment (DCE) for terminals operating in the packet mode and connected to public data networks by dedicated circuit, Revision 1, 1993.
16. Brewster, R. L. 1986. *Data communications and networks*. Peter Peregrinus.
17. ITU-T Recommendation V.24: List of definitions for interchange circuits between data terminal equipment (DTE) and data-circuit-terminating equipment (DCE), Revision 1, 1993.
18. Brewster, R. L. 1986. *Telecommunications technology*. Ellis Horwood.
19. ITU-T Recommendation G.732: Characteristics of primary PCM multiplex equipment operating at 2048 kbit/s.
20. ITU-T Recommendation Q.701: Functional description of the message transfer part (MTP) of signalling system no. 7.
21. ITU-T Recommendation I.412: ISDN user–network interfaces – interface structures and access capabilities, 1988.
22. ITU-T Recommendation I.430: Basic user–network interface – layer 1 specification, Revision 1, 1993.

23. ITU-T Recommendation I.431: Primary rate user–network interface – layer 1 specification, Revision 1, 1993.

24. ITU-T Recommendation I.210: Principles of telecommunication services supported by an ISDN and the means to describe them, Revision 1, 1993.

25. ITU-T Recommendation I.411: ISDN user–network interfaces – reference configurations, Revision 1, 1993.

26. Gasser & Renz 1985. System 12 transmission at 144 kbit/s on digital subscriber loops. *Electrical Communication* 59, no. 1/2.

27. Bellamy, J. C. 1991. *Digital telephony*, 2nd edition, John Wiley (an excellent book).

28. ITU-T Recommendation I.440: ISDN user–network interface data link layer – general apsects, 1988 (now obsolete, refers to Q.920, which is updated).

29. ITU-T Recommendation I.441: ISDN user–network interface data link layer specification, 1988 (now obsolete, refers to Q.921, which is updated).

30. ITU-T Recommendation I.450: ISDN user–network interface layer 3 – general apsects, 1988 (now obsolete, refers to Q.930, which is updated).

31. ITU-T Recommendation I.451: ISDN user–network interface layer 3 specification for basic call control, 1988 (now obsolete, refers to Q.931, which is updated).

32. ITU-T Recommendation I.452: Generic procedures for the control of ISDN supplementary services, 1988 (now obsolete, refers to Q.932, which is updated).

33. ITU-T Recommendation Q.920: Digital subscriber signalling system no. 1 (DSS 1) – ISDN user–network interface data link layer – general apsects, Revision 1, 1993.

34. ITU-T Recommendation Q.921: Digital subscriber signalling system no. 1 (DSS 1) – ISDN user–network interface data link layer specification, Revision 1, 1993.

35. ITU-T Recommendation Q.930: Digital subscriber signalling system no. 1 (DSS 1) – ISDN user–network interface layer 3 – general aspects, Revision 1, 1993.

36. ITU-T Recommendation Q.931: Digital subscriber signalling system no. 1 (DSS 1) – ISDN user–network interface layer 3 specification for basic call control, Revision 1, 1993.

37. ITU-T Recommendation Q.932: Digital subscriber signalling system no. 1 (DSS 1) – Generic procedures for the control of ISDN supplementary services, Revision 1, 1993.

38. ITU-T Recommendation I.410: General aspects and principles relating to recommendations on ISDN user–network interfaces, 1988.

39. Bowker, R. 1988. *Racal Interlan on interoperability*. Racal Interlan: Boxborough, MA.

40. ITU-T Recommendation G.711: Pulse code modulation (PCM) of voice frequencies, 1988.

41. ITU-T Recommendation I.420: Basic user–network interface, 1988.

42. ITU-T Recommendation I.421: Primary rate user–network interface, 1988.

43. ITU-T Recommendation G.704: Synchronous frame structures used at primary and secondary hierarchical levels, Revision 1, 1991.

44. ITU-T Recommendation I.460: Multiplexing, rate adaption and support of existing interfaces, 1988

45. ITU-T Recommendation I.461: Support of X.21, X.21 bis and X.20 bis-based data terminal equipments (DTES) by an integrated services digital network (ISDN), 1988 (now obsolete, refers to X.30, which is updated).

46. ITU-T Recommendation I.462: Support of packet-mode terminal equipment by an ISDN, 1988 (now obsolete, refers to X.31, which is updated).

47. ITU-T Recommendation I.463: Support of data terminal equipments (DTES) with V–series type interfaces by an integrated services digital network (ISDN), 1988 (now obsolete, refers to V.110, which is updated).

48. ITU-T Recommendation I.464: Multiplexing, rate adaption and support of existing interfaces for restricted 64 kbit/s transfer capability, Revision 1, 1991.

49. Proceedings of the ICSC, Paris, 1966.
50. ITU-T Recommendation Q.721: Functional description of signalling system no. 7 telephone user part (TUP), 1988.
51. ITU-T Recommendation Q.722: General function of telephone messages and signals, 1988.
52. ITU-T Recommendation Q.723: Formats and codes, Revision 1, 1993.
53. ITU-T Recommendation Q.724: Signalling procedures, Revision 1, 1993.
54. ITU-T Recommendation Q.725: Signalling system no. 7 – signalling performance in the telephone application, Revision 1, 1993.
55. ITU-T Recommendation Q.741: Signalling system no. 7 – data user part, 1988.
56. ITU-T Recommendation X.61: Signalling system no. 7 – data user part, 1988.
57. ITU-T Recommendation Q.761: Functional description of the ISDN user part of signalling system no. 7, Revision 1, 1993.
58. ITU-T Recommendation Q.762: General function of messages and signals of the ISDN user part of signalling system no. 7, Revision 1, 1993.
59. ITU-T Recommendation Q.763: Formats and codes of the ISDN user part of signalling system no. 7, Revision 1, 1993.
60. ITU-T Recommendation Q.764: Signalling system no. 7 – ISDN user part signalling procedures, Revision 1, 1993.
61. ITU-T Recommendation Q.766: Performance objectives in the integrated services digital network application, Revision 1, 1993.
62. ITU-T Recommendation Q.711: Signalling system no. 7 – functional description of the signalling connection control part, Revision 1, 1993.
63. ITU-T Recommendation Q.712: Definition and function of SCCP messages, Revision 1, 1993.
64. ITU-T Recommendation Q.713: Signalling system no. 7 – SCCP formats and codes, Revision 1, 1993.
65. ITU-T Recommendation Q.714: Signalling system no. 7 – signalling connection control part procedures, Revision 1, 1993.
66. ITU-T Recommendation Q.716: Signalling system no. 7 – signalling connection control part (SCCP) performance, Revision 1, 1993.
67. ITU-T Recommendation Q.767: Application of the ISDN user part of CCITT signalling system no. 7 for international ISDN connections, 1991.
68. Frieberger, P. & M. Swaine. *Fire in the valley.* Osborne/McGraw Hill.
69. Bimpson, A. D. *et al.* (1986) Customer signalling in the ISDN. *British Telecommunication Engineering* 5, April.
70. ITU-T Recommendation Q.933: Digital subscriber signalling system no. 1 (DSS 1) – signalling specification for frame-mode basic call control, 1993.
71. ITU-T Recommendation Q.950: Digital subscriber signalling system no. 1 (DSS 1) – supplementary services protocols, structures and general principles, 1993.
72. ITU-T Recommendation Q.951: Stage 3 description for number identification services using DSS 1, parts published separately in 1992 and 1993.
73. ITU-T Recommendation Q.952: Stage 3 service description for call offering supplementary services using DSS 1 – diversion supplementary services, 1993.
74. ITU-T Recommendation Q.953: Stage 3 description for call completion supplementary services using DSS 1, parts published separately in 1992 and 1993.
75. ITU-T Recommendation Q.954: Stage 3 description for multiparty supplementary services using DSS 1, parts published separately in 1993.
76. ITU-T Recommendation Q.955: Stage 3 description for community of interest supplementary services using digital signalling system no. 1, parts published separately in 1992 and 1993.
77. ITU-T Recommendation Q.957: Stage 3 description for additional information transfer supplementary services using DSS 1, 1993.

78. ISO/DIOS 7498/DAD 1: Information processing systems – open systems interconnection, basic reference model.
79. ITU-T Recommendation X.200: Reference model of open systems interconnection for CCITT applications, 1988.
80. Bartree, T. G. (ed.) 1986. *Digital communication*. New York: Howard Sams.
81. Roca, R. T. 1986. ISDN Architecture. AT&T *Technical Journal* 65, issue 1, January—February.
82. Schlanger, G. G. 1986. An overview of signalling system no. 7. IEEE *Journal on Selected Areas in Communications* SAC-4, no. 3, May.
83. Bowker, R. 1988. *Racal Interlan on interoperability*. Racal Interlan: Boxborough, MA.
84. ITU-T Recommendation I.130: Method for the characterization of telecommunication services supported by an ISDN and network capabilities of an ISDN, 1988.
85. ITU-T Recommendation I.432: B-ISDN user–network interface – physical layer specification, Revision 1, 1993.
86. Cochrane, P. & D. Heatley 1995. Aspects of optical transparency. *British Telecommunications engineering* 14(1).
87. Holliday, C. R. 1995. The national information infrastructure. *Telephony* 24.
88. ITU-T Recommendation I.361: B-ISDN ATM layer specification, Revision 1, 1993.
89. ITU-T Recommendation I.362: B-ISDN adaptation layer (AAL) functional description, Revision 1, 1993.
90. ITU-T Recommendation I.363: B-ISDN adaptation layer (AAL) specification, Revision 2, 1993.
91. ITU-T Recommendation Q.2100: being drafted.
92. ITU-T Recommendation Q.2110: being drafted.
93. ITU-T Recommendation Q.2130: being drafted.
94. ITU-T Recommendation Q.2931: being drafted.
95. Matthews, M. 1991. The synchronous digital hierarchy. Part 1: the origin of the species. IEE *Review* May. Part 2: survival of the fittest. IEE *Review* June.
96. ITU-T Recommendations G.707: Synchronous digital hierarchy bit rates, Revision 2, 1993.
97. ITU-T Recommendation G.709: Synchronous multiplexing structure, Revision 2, 1993.
98. ITU-T Recommendation I.233: Frame-mode bearer services, 1991.
99. ITU-T Recommendation I.501: Frame-mode bearer services interworking, 1993.
100. ITU-T Recommendation I.372: Frame relaying bearer service network–to–network interface requirements, 1993.
101. ITU-T Recommendation I.122: Framework for providing additional packet-mode bearer services, Revision 1, 1993.
102. ITU-T Recommendation I.232: Packet-mode bearer services categories, 1988.
103. ETS 300 324: V interfaces at the digital local exchange (LE) V5.1 interface for the support of access network (AN)
104. ETS 300 347: V interfaces at the digital local exchange (LE) V5.2 interface for the support of access network (AN).
105. ITU-T Recommendation G.804: ATM cell mapping into plesiochronous digital hierarchy (PDH).11/93.
106. Gould, J. ATM's long, strange trip to the mainstream. *Data Communications International* June 1994.
107. ITU-T Recommendation I.413: B-ISDN user–network interface, Revision 1, 1993.
108. ITU-T Recommendation I.321: B-ISDN protocol reference model and its application, 04/91.
109. ITU-T Recommendation I.320: ISDN protocol reference model, Revision 1, 11/93.
110. Wright, T. 1995. An architectural framework for networks. *British Telecommunica-*

tions Engineering 14(2).

111. ITU-T Recommendation G.703: Physical/electrical characteristics of hierarchical digital interfaces, Revision 1, 04/91.

112. Rickard, N. ABR: Realizing the promise of ATM. *Telecommunications*, April 1995.

113. Ballart, R. 1989. SONET: now it's the standard optical network. IEEE *Communications Magazine* March.

114. ITU-T Recommendation G.733: Characteristics of primary PCM multiplexing equipment operating at 1.544 Mbit/s, 1988.

115. ITU-T Recommendation G.702: Digital hierarchy bit rates, 1988.

116. ANSI T1.105–1988. SONET.

117. ETS 300125, equivalent to ITU-T Recommendations Q.920[33] and Q.921[34].

118. ETS 300102–1, equivalent to ITU-T Recommendation Q.930[35] and Q.931[36].

119. Grotemeyer, P. & J. Trömel 1994. VISYON (variable intelligent synchronous optical network), a project for synchronous transmission in the local network. *Phillips Telecommunications Review* 52(2) August.

120. Vincent, T. 1995. The superhighway in action. IEE *Review* 41(3) May.

121. Lindholt Hansen, C. *et al.* 1994. Fibre to the home field trial in Ballerup, Denmark. *Ericsson Review* no. 2.

122. Galvin, M. and A. Hauf 1994. The little engine that could. *Telephony* 19 December.

123. Goddard, M. 1994. The new international telecommunication union. *British Telecommunications Engineering Journal* 13, Part 2, July.

124. CCITT Recommendation X.136: Accuracy and dependability performance values for public data networks when providing international packet-switched services, Revision 1, 1992.

125. ITU-T Recommendation I.330: ISDN numbering and addressing principles, 1988.

126. ITU-T Recommendation E.164: Numbering plan for the ISDN era, Revision 1, 1991.

127. ITU-T Recommendation E.165: Timetable for coordinated implementation of the full capability of the numbering plan for the ISDN era, 1988.

128. OFTEL 1989. *Numbering of telephony services into the 21st century*, July. London: OFTEL.

129. OFTEL 1995. *Geographic telephone numbers, a consultative document*, June. London: OFTEL.

130. Langley, G. & J. Ronayne 1993. *Telecommunications primer*, 4th edn.

131. IEEE 802 family of standards defining data network protocols including Ethernet and FDDI LAN protocols and DQDB MAN protocol, IEEE 802.6.

132. Veitch, P. A. *et al.* 1995. Restoration strategies for future networks. *Electronics and Communication Engineering Journal* 7(3).

133. Lane, J. 1993. Rethinking SONET rings. *Telephony* 4 October 1993.

134. Ramakrishnan, K. K. & P. Newman 1995. ATM flow control: inside the great debate. *Data Communications International* June.

135. T1X1.2, SONET hypothetical reference circuit (HRC), T1X1.2/93–015, March 1993.

136. ITU-T Recommendation G.823: The control of jitter and wander within digital networks which are based on the 2048 kbit/s hierarchy, Revision 1, 1993.

137. ITU-T Recommendation G.783: Characteristics of synchronous digital hierarchy (SDH) multiplexing equipment functional blocks, Revision 1, 1994.

138. Cook, T. 1994. SDH pointer problems. *Telecommunications International* August.

139. Klein, M. & R. Urbansky 1993. Network synchronization – a challenge for SDH/SONET. IEEE *Communications Magazine* September.

140. Eriksen, P. *et al.* 1995. The Apollo demonstrator – new low-cost technologies for optical interconnects. *Ericsson Review* no. 2.

141. Sessions, M. 1995. ISDN – confusion in tariffing, TMA *News and Views* June.

142. ITU-T Recommendation X.28: DTE/DCE interface for a start-stop mode data termi-

nal equipment accessing the packet assembly/disassembly facility (PAD) in a public data network situated in the same country, Revision 1, 1993.

143. ITU-T Recommendation I.250: Definition of supplementary services, 1988.

144. ITU-T Recommendation I.251–257: Stage 1 descriptions of supplementary services.

145. ITU-T Recommendation Q.80: Introduction to stage 2 descriptions for supplementary services, 1988.

146. ITU-T Recommendation Q.81 to Q.87 each in several parts: Stage 2 descriptions for supplementary services.

147. ITU-T Recommendation Q.730: Signalling system no. 7 – ISDN supplementary services, 1993.

148. ITU-T Recommendation Q731–737 each in several parts: Stage 3 descriptions for supplementary services using signalling system no. 7.

149. Powell, S. 1990. Development of Centrex within BT Severnside District Service pbx. *British Telecommunications Engineering* 8(4).

150. ITU-T Recommendation Q.951 to Q.957 each in several parts: Stage 3 descriptions for supplementary services using digital signalling system no. 1.

151. ETS 300195: integrated services digital network (ISDN); supplementary service interactions digital subscriber signalling system no. 1 (DSS 1) protocol, 1995–02.

152. Gross, D. 1995. *Telecommunications, a technology guide.* Financial Times Management Reports: London.

153. ITU-T Recommendation Q.699: Interworking between the digital subscriber signalling system layer 3 protocol and the signalling system no. 7 ISDN user part, 1988.

154. Hoppitt, C. E. & J. W. D. Rawson 1991. The UK trial of fibre in the loop. *British Telecommunications Engineering* 10, Part 1, April.

155. ITU-T Recommendation G.825: The control of jitter and wander within digital networks whch are based on the synchronous digital hierarchy (SDH), 03/93.

156. ITU-T Recommendation G.781: Structure of recommendations on equipment for the synchronous digital hierarchy (SDH), 01/94.

157. ITU-T Recommendation G.782: Types and general characteristics of syncronous digital hierarchy (SDH) equipment, 01/94.

158. Wilmott, C. 1995. *ATM Asynchronous Transfer Mode.* CMI Report.

159. Hawker, I. & P. Cochrane 1995. The "really intelligent network". *British Telecommunication Engineering* 13(4).

160. Beau, O., J. M. Silva, H. Verhille 1990. Network aspects of broadband ISDN. *Electrical Communication* 64(2/3).

161. ITU-T Recommendation X.210: Information technology – Open systems interconnection – Basic reference model – Conventions for the definition of OSI services, Rev 1, 1993.

162. Joel, Amos E. Jr. 1966. *IEEE Journal on Selected Areas in Communications* 14(2).

163. ITU-T Recommendation G.708: network node interface for the synchronous digital hierarchy, Revision 2, 1993.

Further reading

ANSI T1.103–1987. Syntran.

ANSI T1.105–1988. SONET.

Balcer, W. R. 1991. Equipment for SDH networks. *British Telecommunications Engineering* 10, Part 2, July 1991.

Ballart, R. 1989. SONET: now it's the standard optical network. *IEEE Communications Magazine*, March.

Blume, J. *et al.* 1992. Control and operation of SDH network elements. *Ericsson Review* no. 3.

Boehm, R. J. 1989. SONET: the next phase. *Telecommunications*, June.

BT Specification RC8495, First Generation SDH Multiplexer.

Byrne, D. 1991. Don't inconvenience the customers. *TE&M*, 1 November.

Croft, A. 1992. Implementing the national code change. *British Telecommunications Engineering* 10, Part 4.

Daneels, J. & G. Granello 1993. Overview – evolving transport methods in the telecommunications network. *Electrical Communication* no. 4, 1993.

Danielsson, S. 1992. An introduction to the Ericsson transport network architecture. *Ericsson Review* no. 3, SDH.

Debuysscher, P. *et al.* 1990. Evolution towards broadband. *Electrical Communication* 64, no. 2/3.

Depouilly, B. 1990. Overview. *Electrical Communication* 64, no. 2/3 Introduction to issue.

Fleming, S. 1989. To know SONET, know your VTs. *TE&M*, 15 June.

Fontaine, B. *et al.* 1990. Alcatel's view on broadband deployment. *Electrical Communication* 64, no. 2/3.

Gallagher, R. M. 1991. Managing SDH Network Flexibility. *British Telecommunications Engineering* 10, Part 2, July.

Gonzalez Soto, O. *et al.* 1993. SDH Network planing and management. *Electrical Communication* no. 4.

Harrison, K. R. 1991. The new CCITT synchronous digital hierarchy: introduction and overview. *British Telecommunications Engineering* 10, Part 2, July.

Henderson, A. 1988. Into the synchronous era. *Telecommunications*, December.

Henny, D. 1991. The four-digit solution. *TE&M*, 15 August.

Lagerstedt & Nyman 1993. ATM in public telecommunications networks. *Ericsson Review* no. 3.

Lebender, B. *et al.* 1993. SDH Network element technology – the software platform. *Electrical Communication* no. 4.

Luxner, L. 1990. A number crunch: area code crisis plagues Bellcore. *Telephony*, 24 December.

McLeod, N. A. C. 1992. Development of the national code change. *British Telecommuni-*

cations Engineering 10, Part 4, January.

Miller, T. C. 1989. SONET and BISDN: a marriage of technologies. *Telephony*, 15 May 1989.

Powell, W. E. *et al.* 1993. Synchronization and timing of SDH networks. *Electrical Communication* no. 4.

Reid, A. B. D. 1991. Defining network architecture for SDH. *British Telecommunications Engineering* 10, Part 2, July.

Robrock, A. 1989. Test yourself: how much do you know about international communications? IEEE *Communications Magazine* 27, no. 12, Dec.

Sexton, M. *et al.* 1993. SDH architecture and standards. *Electrical Communication* no. 4.

Van Bogaert, J. *et al.* 1993. SDH network element technology – the hardware platform. *Electrical Communication* no. 4.

Wright, T. C. 1991. SDH multiplexing concepts and methods. *British Telecommunications Engineering* 10, Part 2, July.

Index